Territorial Crisis Management

SCIENCES

Geography and Demography, Field Director – Denise Pumain

Geography of Risk, Subject Head – Samuel Rufat

Territorial Crisis Management

From Emergency to Reconstruction

Coordinated by
Richard Laganier
Yvette Veyret

iSTE

WILEY

First published 2022 in Great Britain and the United States by ISTE Ltd and John Wiley & Sons, Inc.

ISTE Ltd
27-37 St George's Road
London SW19 4EU
UK

www.iste.co.uk

John Wiley & Sons, Inc.
111 River Street
Hoboken, NJ 07030
USA

www.wiley.com

Library of Congress Control Number: 2022939248

British Library Cataloguing-in-Publication Data
A CIP record for this book is available from the British Library
ISBN 978-1-78945-080-4

ERC codes:
SH2 Institutions, Values, Environment and Space
SH2_1 Political systems, governance
SH2_5 International relations, global and transnational governance
SH2_11 Human, economic and social geography

Contents

Foreword ix

Samuel RUFAT

Introduction xiii

Richard LAGANIER

Chapter 1. Territorial Crisis, Elements of Definition 1

Richard LAGANIER and Yvette VEYRET

1.1. Crisis and catastrophe 6
1.2. Disasters of natural origin: a circumscribed crisis territory 10
1.3. Localized crises of natural origin aggravated or provoked by human activities 16
1.4. Industrial, technological and nuclear disasters and crises: localized crises and spatial consequences 17
1.5. More diffuse environmental, food and health crises on a regional or continental scale 21
1.6. “Reticular” crises 24
1.7. The “mega-crises” on a global scale, the domino effects 27
1.8. Conclusion 30

Chapter 2. General Principles of Crisis Management 31

Freddy VINET and Anne-Marie LEVRAUT

2.1. Nature of the crisis and forms of crisis management 33
2.1.1. The nature of the crisis: cycle, trajectory or “black swan”? 33
2.1.2. The nature of the crisis: scale of the crisis versus scale of its management 34
2.2. Preparing for the crisis: the prevention/preparation link 37

2.2.1. Preparation: the imperative link between prevention and crisis management 37
2.2.2. Monitoring: choosing the right indicators. 39
2.2.3. From monitoring to forecasting 39
2.2.4. Forecasting tools 40
2.3. Information in times of crisis 41
2.3.1. Upstream information in times of crisis: monitoring tools 42
2.3.2. Alert 43
2.3.3. Infodemia, misinformation and controversy 46
2.4. Crisis management: an integrated approach. 47
2.4.1. Conditions for a shared vision 47
2.4.2. Integration of populations 49
2.4.3. The importance of logistics 53
2.4.4. Population management: the issue of evacuation 54
2.5. Conclusion 55

Chapter 3. Learning from Crisis Management 57

Freddy VINET, Alain CHEVALLIER, Hoilid LAMSSALAK and Dimitri LAPIERRE

3.1. Feedback: objectives, contents, actors 60
3.1.1. Objectives of feedback 61
3.1.2. Feedback times: feedback and territories 62
3.1.3. Feedback methods and contents 65
3.1.4. Feedback: a tool for sharing and disseminating learning 66
3.1.5. Conclusions on feedback. 69
3.2. Crisis management exercises and simulation 69
3.2.1. Objectives and prerequisites 69
3.2.2. Organizing a crisis management exercise: methodology 71
3.2.3. Types of exercise 74
3.2.4. Communication and media awareness 75
3.2.5. Stress management 76
3.2.6. A community-wide exercise 77
3.3. Conclusion 79

Chapter 4. Crisis Management Resources 81

Jérémy ROBERT, Pauline GLUSKI, Sébastien HARDY, Pascale METZGER and Alexis SIERRA

4.1. The reason for a shift from risk to crisis 83
4.1.1. Why be interested in crises? 83
4.1.2. From stakes to resources: conceptual framework 84
4.2. How to identify crisis management resources? 87
4.2.1. Different types of crisis management resources 87
4.2.2. Methodological aspects. 94

4.3. The benefits of a resource-based approach 105
4.3.1. Understanding the vulnerability of the territory 105
4.3.2. An operational interest 106
4.4. Conclusion: thinking about a "risk–crisis–development" continuum 107

Chapter 5. Post-disaster Recovery: Challenges and Resources 109

Annabelle MOATTY

5.1. The challenge of coordinating a multitude of actors with inadequate regulatory frameworks 111
5.1.1. The actors and temporalities of the reconstruction process 111
5.1.2. Legislative and regulatory frameworks for recovery. 115
5.2. Financial resources for reconstruction and the weight of solidarity in individual recovery 117
5.2.1. Government resources 117
5.2.2. International aid directed mainly to developing countries. 119
5.2.3. The influence of solidarity mechanisms on the recovery of individuals and communities 120
5.3. Land resources and territorial restructuring 122
5.3.1. Relocation of areas deemed too dangerous for population settlement . . . 122
5.3.2. On-site reconstruction, in the continuity of pre-existing structures. 125
5.3.3. Post-disaster as a catalyst for inequality? 126
5.4. Conclusion 128

Chapter 6. Crises and Territories: Legacies, Inertia and Dynamics 131

Nancy MESCHINET DE RICHEMOND and Yvette VEYRET

6.1. Crises and territories: always complex articulations, in constant renewal . . . 132
6.1.1. Crises and territories: scales, actors and contexts 132
6.1.2. Obstacles linked to conflicts and the interplay of actors in crisis management. 137
6.1.3. Crises and urban territories: rupture, bifurcation, sustainability, resilience? 143
6.2. Inertia of representations 145
6.2.1. Inertia of representations on the role of the population and its vulnerabilities 146
6.2.2. Inertia of the representations of the crisis itself 149
6.2.3. Territorial inertia feeds crises: a territorialization of data unsuited to management? 154
6.3. Crises feed territorial inertia 156
6.3.1. When crises and emergency become everyday life: the example of Haiti 157

6.3.2. Instrumentalization and resilience: the root causes of crises are still present 158

Chapter 7. Founding Crisis of Territorial Renewal 165

Yvette VEYRET and Richard LAGANIER

7.1. Managing a risk territory on a global scale 168
7.1.1. The age of risk management 169
7.1.2. SYMADREM 170
7.1.3. The Rhône Plan, a global vision 170
7.1.4. Crisis management 172
7.1.5. Evolution of the legislation, the future of SYMADREM 173
7.2. Reducing risk through protection 174
7.2.1. Dikes and diking systems, regulation, management 175
7.2.2. The equipment of rivers with polders 177
7.2.3. Soft management of coastal risk areas 179
7.3. Adapting the habitat 182
7.3.1. Responses to natural hazards 183
7.3.2. Housing and technological risks 186
7.3.3. Adapting buildings to the effects of climate change 188
7.4. Adapting the neighborhood and the city 189
7.4.1. Flood-safe neighborhoods 189
7.4.2. Urban technical networks and resilient critical infrastructures 194
7.4.3. Responses to climate risks – the urban heat island, "green" neighborhoods 195
7.4.4. Eco-neighborhoods, resilient neighborhoods? 196
7.4.5. From the eco-neighborhood to the "smart city" 198
7.5. Strategic retreat 200
7.6. Conclusion 203

References 205

List of Authors 233

Index 235

Foreword

Samuel RUFAT
Institut Universitaire de France, Paris, France

In France, in 2021, the Council of State pointed out that after the 2015 attacks and the 2020 Covid-19 pandemic, France has spent more than half of the last six years under a state of emergency. Magistrates are concerned about the recent inclusion of emergency measures in common law. However, these are not the first attacks, nor the first pandemic. As Albert Camus already pointed out: "pestilence is in fact very common, but we find it hard to believe in a pestilence when it descends upon us. There have been as many plagues in the world as there have been wars, yet plagues and wars always find people equally unprepared" (1947, *The Plague*). Crises reveal the vulnerabilities of our societies, often overwhelming planning and anticipation, imposing uncertainty in decision-making, the difficulty of committing sufficient resources in advance and the necessary adaptation to the degraded functioning of territories, sometimes over the long term. Are we witnessing an increase in the number of environmental, climatic, social and economic crises that increasingly appear to be an obstacle to the sustainability of territorial development? In return, should we rethink the functioning and management of territories in order to make uncertainty, degraded functioning, even improvisation, the generalization of emergency and exceptional measures the new normality?

In this subject, "Geography of Risk", works take stock of the renewal of notions, approaches, issues and tools at the articulation between risks, societies and the environment, because risks do not always translate into crises or disasters and the

Territorial Crisis Management,
coordinated by Richard LAGANIER and Yyvette VEYRET.

increase of risks is not inevitable. It is possible to reduce the exposure of populations and their vulnerabilities, and support their adaptation or the resilience of societies in the face of risks that cannot be reduced, or the effects of crises that cannot be distributed. Risk prevention, disaster reduction and crisis mitigation require cutting edge approaches, in a subtle balance between the social demand for coordinated action at all scales and the imperative of reflexive criticism, in order to address the substance of the problems, rather than the most visible symptoms or the most fashionable solutions. This book shows that, as for other concepts, "scientific production in this field of crisis management remains limited, particularly in the field of geography, and does not seem to meet the needs of managers, which are often still ill-defined".

Everything in the media and political discourse becomes a crisis, an emergency or an uncertainty. This book shows that it is because "our societies seem to be becoming more and more crisis-prone" that it is essential to propose a critique of the staging of crises, and question the heuristics of crisis thinking and the effects of its management on territories. Richard Laganier and Yvette Veyret have brought together a multidisciplinary team of French-speaking specialists, researchers and academics in order to put their expertise into dialogue. The authors offer a critical analysis of the links between risks, crises, disasters and territorial development, without glossing over the political and ethical issues, as well as the practical difficulties. They show in a pedagogical way the conceptual, practical and strategic questions under the great diversity of crises, always with the aim of drawing lessons from experience and proposing ways to improve crisis and territorial management: "crises are not only dysfunctions and disturbances, they are also moments of rupture in the organization and functioning of territorial systems, ruptures that can lead to a different way of thinking about post-crisis preparation and reconstruction."

In this book, Richard Laganier, Yvette Veyret and their team draw up a critical assessment of the issues, lessons learned from a wide variety of case studies on all continents, uncertainties, scale and collateral effects, while suggesting that social scientists and managers adopt a renewed look at the links between crises and territories. On the one hand, they propose to consider the crisis "as an opportunity to rethink the development of a society in order to make it less vulnerable, more equitable and to implement a more sustainable use of resources". On the other hand, they warn that "the crisis can feed territorial inertia, providing emergency solutions at certain scales, while at other scales, its management contributes to the maintenance of vulnerable or ill-adapted organizational frameworks, but also to the permanence of inequalities and power relations between actors, territories or countries". They stimulate an essential reflection to support societies and territories in the face of the global climate crisis, environmental degradation, with an emphasis on emergencies, in the perspective of a sustainable and livable future for all.

Introduction

Richard LAGANIER
Université de Paris, France

Societies, which are now largely globalized and almost 60% urban, have become very tense. Due to high human concentrations, multiple mobilities and flows on a global scale, interrelationships and interdependencies, as well as the civilizational, cultural, political, economic or religious tensions that result from them, they carry within them the seeds of crises of various origins.

These last few decades have been struck by multiple crises, some of them limited to portions of territories (September 11, 2001 attack in New York, September 21, 2001 AZF industrial disaster in Toulouse) but with indirect impacts that go far beyond the areas immediately affected. Others are immediately global in scope (climate change and its impacts; Covid-19 health crisis in 2020–2021) to the point that they can sometimes be perceived, rightly or wrongly, as imported crises of distant or unknown origin. The intertwining of vital infrastructures (energy, water, communication), which is increasingly characteristic of our societies, is also an aggravating factor, both in terms of the spread of the crisis through domino effects (earthquake in Japan, which led to a tsunami and a nuclear accident in Fukushima on March 11, 2011), and in terms of the often complex and multidimensional management of the crisis. Moreover, some of these crises can destabilize because of their speed, power and the surprise effect induced (an attack, a very rare and large-scale natural hazard), while a large part of the population finds it difficult to consider others, which are more diffused and slower (global environmental changes), as crises.

However, all of these crises have many points in common. They reveal many underlying fragilities and vulnerabilities of our societies. They underline the complexity of territorial systems and thus the organizational challenge of dealing with catastrophic events. They threaten the values that structure the foundation of our societies and even the very functioning of the territorial and political systems that organize life. Crises question the decision-making processes and put the decision-makers who have to act in an uncertain environment under pressure, due to a lack of information or resources, to ensure, despite everything, the functioning of the territorial system in degraded mode. However, crises do not respect the crisis management plans prepared in advance. Each time, they constitute an organizational challenge that requires innovation, rapid reconsideration of action methods and adaptation to deal with the many unknowns that mark out the crisis.

Because our societies are increasingly crisis-prone, it seems necessary, not to say urgent, to explore these crises and the associated practices of anticipation, management and post-crisis, and to think of a risk–crisis–territorial development continuum. In this perspective, this book proposes both conceptual and strategic questioning. Conceptual because it allows us to better define a notion that is widely used today in many spheres of society and whose contours need to be clarified (Chapter 1). Strategic because it aims to clarify and make the modalities of action in the field of crisis management robust (Chapter 2), adjust the practices of those in charge of managing crises and territories and finally, reinforce the learning of decision-makers and citizens to better face future crises (Chapter 3). It is strategic because it also invites us to analyze the resources on which crisis management is based (Chapter 4), as well as those that can be mobilized to think about and organize post-crisis reconstruction (Chapter 5). But societies are not equal in terms of financial, land, cognitive or political resources for integrating crises and drawing lessons from the past in order to design future territorial strategies. Thus, we often see a form of territorial inertia expressed after crises, which should be analyzed (Chapter 6). Others, on the contrary, seek to innovate, imagine renewed territories, integrate preparation for reconstruction into the pre-crisis period and recompose territories now, in order to make them more resilient. The crisis can then be the driving force of these changes and contribute to the emergence of new practices, or even new urban and territorial utopias (Chapter 7).

1

Territorial Crisis, Elements of Definition

Richard LAGANIER[1] and Yvette VEYRET[2]

[1] *Université de Paris, France*

[2] *Université Paris Ouest Nanterre, France*

An earthquake in Haiti, a tsunami in Indonesia, hunger riots, terrorist actions, the Covid-19 crisis and so on, in the media, as well as in political speeches, everything is in crisis. The term *crisis* is borrowed, through the Latin *crisis*, from the Greek *krisis*, which first meant the action or ability to choose before covering the idea of an "accident of a medical nature, sudden and unexpected". For the French language dictionary, it is essentially this last aspect that is preserved, as well as its figurative uses, to designate "a sudden event that comes as a sudden alteration of health, disturbing and upsetting a hitherto peaceful situation". The dictionary of geography defines crisis as "a rupture, or reversal of tendency in an evolution". It can be a rapid, violent manifestation, a worrying situation, a lack of something. We can consider that crisis is manifested by important modifications to the environment (taken in the broadest sense: natural, economic, social, etc.) in which the individual or the group lives, modifications responsible for worries, uneasiness, insecurity or even threats to the lives of the individuals and death. Crisis requires decisions and treatment for maximum resilience. Morin and Béjin (1976) insist on the idea of decision: "it is the decisive moment in the evolution of an uncertain process, which allows the diagnosis to be made." But these authors also emphasize that today, crisis means indecision: "it is the moment when, at the same time as a disturbance, uncertainties arise" (Morin and Béjin 1976). If we follow Lagadec (1991), crisis is

Territorial Crisis Management,
coordinated by Richard LAGANIER and Yyvette VEYRET.

"a situation in which multiple organizations, faced with critical problems, subject to strong external pressures and bitter internal tensions, find themselves suddenly and for a long period of time, at the front of the stage, projected against each other". Libaert et al. (2018) insist on the unexpected event at the origin of the crisis, which requires a response, decisions for its management. The work of these different authors is similar and we can, following Crocq et al. (2009) define crisis as a:

> crucial moment in the life of individuals, groups and populations, which marks a break in continuity and uncertainty about the evolution of events, which involves a threat to the values, objectives and functioning of these individuals and groups and which discovers major stakes for their freedom, their integrity, or even their survival; crisis implies the need to act urgently and possibly in a deteriorated situation, it can lead to the return of the previous state, the establishment of a new state of equilibrium or the worsening of the deteriorated state.

A crisis is therefore the consequence of a major event of natural or anthropogenic origin (in the broadest sense), both a process and a state, whose effects affect a large number of people and cause significant damage that may exceed society's capacity to respond.

Morin attempted to establish a theory of crisis, which he named "*crisology*" (Dorna and Costa 2015) whose three basic principles reflect complexity: systemic, cybernetic and negentropic. The systemic principle is related to the complexity of the structures underlying any organization characterized by the attraction and connection existing between the components of the system. The cybernetic principle attributes the maintenance of the stability and coherence of a system to feedback. Feedback, in the broadest sense, is the return action of an effect on the device that gave rise to it, and thus on itself. Negative feedback, triggered by the variation of an element, tends to seek to cancel this variation in order to re-establish the stability and integrity of the system, whereas positive feedback corresponds to a "deviation that increases as it feeds its own development [...] If nothing comes to inhibit or cancel it, the feedback develops in a chain throughout the system. The principle of entropy (the opposite of negentropy) implies that every system is heading towards a natural degradation." If no action is taken, entropy will increase, and this is the crisis (Morin 1976; Masclet 2010). "Crisology" allows us to emphasize the importance and variety of crises inherent to life and to our societies. The risk society envisaged by Beck (2001) is naturally also one of crises that can affect the individual, the group, society and, in the broadest sense, the environment, as shown by the example of the 2020 pandemic, whose effects are sanitary, psychological, economic, social and political.

The analysis of crisis(es) is carried out in many works (Pauchant 1988; Guillaumin 2004). If in the Middle Ages, the term crisis was only used in a medical sense, in the 17th century, on the contrary, Furetière added "a figurative sense, relating to human actions (intrigue, trial) [...] it is from this figurative sense that, according to Walther von Wartburg, the term 'crisis' was transposed from the medical domain to the economic domain, thanks to the difficult economic situation at the end of the 17th century" (Morsel 1988). From the 19th century onwards, the term crisis was used to describe cultural, political and economic changes (Morin op. cit.) and any situation characterized by disturbances and associated uncertainties, whatever its nature, was then defined as a crisis (Revault d'Allonnes 2013). Crisis is present in the works of Malthus, in those of Ricardo in the 18th century, and for 19th century economists, crises appear as inescapable.

> There are no remedies against crises in the sense that crises are inevitable. No measure can avert them; it is at most if one can attenuate their effects, shorten their duration, calm the panic that is inseparable from them. A crisis is an accident in the economic life of people, as it is in the life of individuals: it is feverish activity (Jourdan 1882 cited by Gallois 2012).

Historical works have taken up this notion of crisis. The first theories were born in the middle of the 19th century joining in this the analyses of economists (see Juglar, Marx, etc.). As Morsel (1988) points out, "the use of the notion of 'crisis' in history depends very closely on the historiographic or even epistemological stage of the moment, and probably also, to a degree that is difficult to appreciate, on the context in which the historian is immersed". The historian Perroy recalled that "the word 'crisis' has been used to designate two distinct phenomena. Sometimes it is about sudden depressions, limited in time […] sometimes a movement of durable and prolonged collapse of the economy" (Morsel op.cit.). Psychology insists on the subjective dimension of crises associated with ruptures in human relationships, "with the confrontation of antagonistic forces; with imaginary stupefaction and difficulties in making decisions" (Barus-Michel et al. 1996). Thus, Morin can underline that "the notion of crisis has spread in the twentieth century to all horizons of contemporary consciousness [...] But this notion, by generalizing, has been emptied from the inside". According to the author, "The word crisis is used to name the unnameable; it refers to a double gap; the gap in our knowledge (at the very heart of the term crisis), the gap in the social reality where the crisis appears" (Morin 1976).

The very broad use of the term crisis sometimes leads to a lack of readability and to the questioning of the use of this term itself, with crisis being considered as inherent to all living systems. How far should we go in its use? Is any change, any modification in relation to a previous situation a crisis? The idea of disturbance or rupture associated with crisis is similar to the deterministic analysis developed by certain ecologists (Duvignaud 1974), according to whom any disturbance or crisis is harmful to an organism or a system, even though these disturbances or crises are intrinsically associated with the evolution of these systems. The proponents of such fixist conceptions defend the idea of an earlier "equilibrium", of a lost golden age to be rediscovered, of the climax and so on. The idea of crisis has also been used in a cyclic conception by historians (Toynbee) or geographers (cycles of erosion, etc.) who rely on the ternary notion of cycle, any system being characterized by three phases of "life": youth, apogee and decline in which the crisis occurs at the time of the passage of the apogee to the decline or designates the whole of the period of decadence. In fact, the notion of crisis must be understood as a change, not implying a return to a time or a point zero, but associated with bifurcations towards new forms of organization or ruptures implying new "departures", with forms of resilience allowing the system to function even on other bases and in a degraded mode.

How do we classify crises? A crisis is always a unique situation to which specific actions correspond (Pauchant and Douville 1993). To facilitate their implementation, the literature presents certain classifications of crises according to their triggers (Mitroff et al. 1987; Roux-Dufort 2003; Westphalen 1992 in Libaert 2018), the internal or external nature of the crisis (Lagadec 1991) or the "targets" concerned: the product, industrial and institutional aspects (Ogrizek et al. 1997). The identification and classification of crises is thus intended to provide managers with a reference framework for the analysis (Lagadec 1995) of these situations, including when the phenomenon is uncertain and evolving (Perrow 1984; Thiétart et al. 1997). Among the ways of considering and dealing with crises (Aluson et al. 1999), the literature offers two approaches: the more event-based approach (Hermann 1963) and the more process-based approach (Forgues 1996). Many authors insist on the importance of threats that generate disruptions and ruptures in organizations or systems and concern the security, safety, integrity of the system and people, in other words, aspects that concern the vital systems of society. The necessary decision-making must take place in a context of uncertainty. Crises also refer to the perception of the different actors who are affected by them (Laganier 2013).

We will consider crises according to the process or processes that justify them, crises related to natural hazards, industrial and technological crises: climate crisis, biodiversity crisis, seismic crisis or industrial accident, and also economic, geopolitical, social and health crises, excluding open conflicts and wars. The geographical extension of crises constitutes another criterion of distinction: the crisis can concern a district, city, region or even the planet. The duration of crises, whether brief (a few days to a few weeks) or much longer (several years, for example), also makes it possible to distinguish them. The types of impact lead us to consider several types of crises: environmental, economic, social and/or political, and also complex crises combining these different aspects. Finally, we can distinguish between crises of unequal importance and highlight "mega-crises" that concern the entire planet, a large number of political and economic actors and the population. The climate crisis, for example, is the result of changes in the climate system and the climate machine on a planetary scale, linked to the quantities of greenhouse gases released by societies into the atmosphere since the Industrial Revolution. This crisis reflects the changes that have taken place over a fairly long period of time in economic practices, transport and, more broadly, lifestyles (industrialization, urbanization, tourism, etc.). It is revealed by numerous and/or more intense manifestations than in the past: heat waves, repeated cyclones, retreat of the summer ice pack in the Arctic Ocean. Some episodes of high intensity are the cause of repeated disasters (storms of 1999, storm Xynthia, cyclone Irma in France). Thus, global crisis can give rise to short-lived events (storms, cyclones, etc.) that are sometimes repetitive, sometimes seasonal (retreat of the ice pack) or more random, all of which nevertheless reinforce the effects of the climate crisis, which also manifests itself over longer periods of time (rise in sea level, accelerated seasonal melting of permafrost, for example).

From what thresholds (economic, social, cultural, etc.) can we speak of crisis? And do we not apply this term to situations of different magnitude and nature? Moreover, the perception of a crisis situation varies a lot according to the individuals. What can appear dangerous and unbearable to some is not for others who accept the danger for religious reasons, for example, or because no other solution can allow them to defend themselves. In other words, is it possible to envisage an objective reality of crisis and on the basis of what criteria? All the more so as one must insist on the importance of the media in "creating a crisis". The media do not simply report on the crisis, but contribute to the "creation" of this event, by giving it the status of a crisis or a catastrophe. The crisis becomes, through the associated media coverage, a kind of spectacle, a theater. "The theater does not have a monopoly on tragedy: radios, televisions and newspapers are the permanent reflection of contemporary tragedies" (Steinle-Feuerbach 1995). Media coverage can help to find solutions to the crisis, while in some cases, its total absence contributes to the forgetting of dramatic events. Since the Armero disaster, we know that "putting a face – that of the little girl gradually swallowed up by the lahar – on this disaster has made it possible to show its magnitude, and has given rise to a great surge of international solidarity" (Arboit 2007). Nevertheless, if communicating in case of crisis is a necessity, in particular to reassure the population, this communication must be transparent, not hide the facts, not modify them for ideological purposes, all of which implies presenting the different useful expertises.

According to the origins of the crisis, its duration, its intensity and its consequences, we can distinguish crises and catastrophes, between which, there exists a “radical difference” (Thom 1976).

1.1. Crisis and catastrophe

The term catastrophe comes from the Latin catastropha, which itself comes from the Greek, and means “reversal”. Catastrophe is in essence a very visible phenomenon, an observable discontinuity and a rupture (Thom 1976). “There is an obvious link between crisis and catastrophe, crisis is often a precursor of catastrophe that it precedes or provokes” (Thom op.cit.). Catastrophe is a product of a social order. Any catastrophe taking place in a given social context is often the revelation of a crisis situation. Catastrophe can cause or at least aggravate a crisis, political or economic (Revet 2007; Cabane et al. 2015). Nevertheless, this distinction is not unanimously accepted, so what we refer to as a disaster is sometimes defined as a crisis, as Le Bouëdec (2011) does in his historical works for which “crisis is a short social time marked by natural accidents, economic setbacks that weaken society. Crisis is a mirror to discover the organization and functioning of the community at the heart of a model”. The author contrasts crisis with the more complex and longer lasting rupture that translates the passage from one economic, social “model” to another. However, in most cases, a disaster is defined as the result of a sudden event causing damage to property, people and ecosystems. Disaster also has a psychological dimension (loss of reference points, social links). In fact, if certain processes at the origin of a disaster are rapid or even almost instantaneous (consequences of a cyclone, an earthquake, an explosion, etc.), in many cases, the process at stake has a certain temporal dimension (droughts, for example) and is more widely part of a crisis situation of variable scale. Disaster results from the unfolding of a hazard, and is a source of danger for populations. But disaster is also the result of an accumulation of multiple vulnerabilities (organizational, material, cultural) that hazards reveal over time. Several criteria make it possible to define a disaster, in particular its origin: a natural or technological event and so on, but it is above all a social event; disaster amounts to transform a natural hazard into a social construction. Without human society, there would be no disaster. Berque (2012) emphasizes that “natural phenomena, whatever their scale (from a shock of particles to a shock of galaxies), are not disasters”. In Fukushima, “on the other hand, it is a disaster, that is to say, something that implies human existence [...] The disasters following the tsunami of March 11, 2011 [...] have as their primary cause the mode of the relationship that modernity has established between Japanese society and its environment, in the oblivion of history and the abstraction of the human subject from its environment, reduced to a clean slate defined by profit alone" (Berque 2012). Disaster is the result of a negative event, even though some actors may benefit from it; it is an extraordinary, exceptional event. There is a disaster when the system concerned does not have the capacity to absorb the event; this technical, cultural, mental and economic capacity, which proves to be insufficient, translates into a weak resilience and, in many cases, a crisis situation. Thus, disaster in a certain number of cases is the revelation of a deeper crisis; a land movement (landslide, rock fall, collapse) responsible for victims and degradation can be a marker of the slow dynamics of a slope often not perceived by the populations concerned, dynamics itself an indicator of deforestation or poorly conducted development of this space. The hazard that manifests itself from a certain threshold of moisture in the rocks and soils can cause a disaster affecting developments and populations. The earthquake in Haiti in 2010 revealed by the magnitude of its effects, more than any other dramatic event (cyclones in particular), the dysfunctions, the socio-political crisis affecting the society of this developing country (corruption, inorganization, poverty, embezzlement of money for the benefit of the leaders alone, mafia system, violence… questionable action of NGOs). To be effective, repairing the effects of the earthquake and the crisis would require discussing and reviewing the functioning of the Haitian political and economic system, which is marked by social and environmental injustice. This is far from being the case, and it is therefore understandable that 10 years after the disaster, the effects of the earthquake are still present. In these examples, we must therefore distinguish between the existence of a multiform crisis generally involving long periods of time, multiple interactions, a complex system and, within the crisis, a disaster that is generally of a shorter duration due to a well-identified triggering process, in this case an earthquake, the effects of which can only aggravate the prior crisis. The example of the Kobe earthquake in 1995, which destroyed the great Japanese port and caused many victims, shows the role of the urbanization of the city of Kobe itself, the price of land, the extreme densification of urban space, and the types of precarious constructions filling the gaps in the urbanization. All of this explains the importance of the fires which broke out during the earthquake and caused the greatest number of victims (more than 5,000 dead and tens of thousands injured) and damage (105,000 homes destroyed and 145,000 damaged). In all these examples, reconstruction should require a broader analysis of the disaster in order to consider what contributed to the extent of the damage of all kinds, beyond the triggering event itself, which is the source of vulnerability. In many cases, the natural or technological event responsible for the disaster is only the emerging part of a more complex situation, of dysfunctions affecting multiple domains (social, economic, etc.). Thus, we can consider the process that triggers the disaster as the immediate cause of the situation, and the crisis in which the disaster occurs as the result of broader, more "distant" factors, in the same way that historians distinguished between immediate and distant causes in the analysis of wars. In many cases, the hazard may appear to be directly responsible for the greater or lesser number of victims and most of the damage, but these are to be put in relation to the vulnerability of the populations, which is itself a function of their economic and social situation, and of their culture, which could potentially bring about the existence of a crisis. "It is enough," writes Rieu (2012) about Fukushima, "to pull the threads of an affair or a disaster so that whole sections of a society are revealed". This observation demonstrates the difficulty of proposing a classification of crises and disasters, which is important to include in a systemic analysis.

When should we use the term disaster? The UN defines a disaster as a process that results in death or loss of life and material damage. A "significant disaster" is characterized by losses of at least 100 human lives and 1% of GDP. The term "cataclysm" refers to events responsible for more than 10,000 victims. Dauphiné (2000) proposed a synthetic scale based on three criteria, the number of victims (from 100 to 9999, the most discriminating factor), financial losses and biomass losses. The database of the *Centre de Recherche sur l'Epidémiologie des Désastres* (CRED 2010) defines a catastrophic event as one that has caused at least 10 deaths, affected 100 people, or required international assistance. The Reinsurance Company Swiss Re uses the term "catastrophe" when the event causes economic losses of at least 85.4 million dollars, more than 20 dead or missing, 50 injured and 2000 homeless. Munich Ré distinguishes among the six degrees of severity of large-scale disaster when the victims are several thousand. In France, if we follow the classification established by the specialized environmental inspection mission, the effects of storms Martin and Lothar in 1999, then Xynthia in 2010 fall into the category of "catastrophe", while the 2020 pandemic is part of the "major catastrophes" category.

Category	Human damage	Property damage
Incident	No injuries	Less than 0.3 million euros
Accident	One or more injured	Between 0.3 and 3 M
Serious accident	1–9 deaths	Between 3 and 30
Very serious accident	9–99 deaths	From 30 to 300
Disaster	100–999 deaths	Between 300 and 3 G
Major disaster	More than 1,000 deaths	3 G or more

Table 1.1. *Damage severity scale (sources: Mission d'inspection spécialisée de l'environnement (MISE), 1999)*

These categorizations have certain limitations. They do not take into account the socio-political consequences of events, which are difficult to measure and sometimes do not become apparent until later (Dauphine 2000), nor do they take into account damage to the environment and/or to heritage.

All natural or technological events do not necessarily fall into the category of disaster, if the number of victims is reduced or even non-existent, and the damage relatively limited. Nevertheless, for the individual or the population who have experienced a disturbing event, generally unforeseen or of very high intensity, and who have lost part of their possessions and memories, the perception of the event is that of a disaster. Thus, in Rouen during the accident at the Lubrizol factory in 2019, the expression "catastrophe" (industrial, economic, ecological) was very quickly used by the media, as well as by the very worried population who even compared the Lubrizol fire to the disasters in Toulouse, Chernobyl and Bhopal.

In France, the term "catastrophe" is widely used. Damage suffered by individuals during a natural event (floods, storms, etc.) can be covered by the "natural disaster" decree. The insurance system instituted by the law of July 13, 1982 relating to the compensation of victims of natural disasters is based on the obligation to include a guarantee against "uninsurable direct material damage caused by the abnormal intensity of a natural agent" in all property damage and business interruption insurance contracts. When an "exceptional" event occurs, the coverage is subject to the publication of a natural disaster decree ("catnat") which specifies the municipalities and periods where the phenomenon and its characteristics occurred. This decree is issued following the opinion of an interministerial commission referred to by the prefect at the request of the commune. This commission recognizes, on the basis of the scientific report, the exceptional character of the natural phenomenon that caused the damage.

The French Institute for the Environment (*Institut français de l'environnement*, IFEN) points out that three quarters of French communes have been subject to at least one "catnat" decree since the law was promulgated. These decrees have concerned municipalities in flood-prone valleys, on the Mediterranean coast, and also on the plateaus of northern France, where mudflows and/or land movements related to mines or underground quarries occur. The effects of the droughts which affect sedimentary France (Paris Basin, Provence, Aquitaine) are at the origin of processes of shrinkage/swelling of clays and of considerable damage to the habitat.

The cost of disasters is difficult to measure. Nussbaum (in Veyret et al. 2013) points out that over the last 25 years, 36 major events have resulted in compensation of more than 100 million euros, their total cost is estimated at more than 35 billion euros, a third of which in 2011 alone, the human cost is 406 deaths, 129 of which were in 1999 (storm Martin and Lothard). Floods are the most frequent disasters, but storms are the most deadly. The increasing costs of disasters are obvious despite the importance of risk management and crisis prevention. However, the concentration of the population and various developments in areas at risk is also a reality (coastal development, urban expansion along rivers, etc.). In addition, the worsening of high intensity phenomena in relation to global warming contributes to the multiplication of "catnat" decrees.

Finally, the distinction between crisis and disaster is sometimes difficult to make, as one term can replace the other. In France, institutional actors widely use the term crisis, including to designate the consequences of natural hazards or industrial dangers at the origin of an incident and/or disaster. The state and its deconcentrated organizations also use the term crisis when it comes to managing a dangerous event at the national and regional levels, while mayors are responsible for crisis management at the communal level. The term crisis is often used both to consider what is more akin to a disaster and to define the political, social, economic and health situation that creates vulnerabilities, which we have defined as a crisis in the strict sense.

Can we consider the modalities of the unfolding of a crisis often defined as a cycle characterized by several phases or stages? A large number of authors have described different stages in the development of a crisis, the preliminary phase that allows precursory signs to appear, more or less recognized according to their nature and the level of preparation. During the next phase, the signs of the crisis or disaster appear, to which the organizations concerned respond in the third phase, which must contribute to limiting the damage. The last phase is that of the end of the crisis, but the crisis may remain present through the material traces that persist (change in the use of a space affected by the crisis, expropriation, etc.), in the collective memory as well as in that of individuals (Mitroff et al. 1987; Fink 2000). This last phase is not a return to the initial precrisis situation. Crises and disasters are not all of the same magnitude; they will be considered first by taking into account the nature of the triggering processes, the space and the associated time.

1.2. Disasters of natural origin: a circumscribed crisis territory

Hazards (avalanches, floods, storms, earthquakes, etc.) occur in a specific area, which may be a portion of a commune (in the case of a very localized fire hazard or landslide), in several communes, or in a linear pattern that extends over several communes or even several regions in the case of floods. The hazard can also affect a large area, as was the case with the storms that swept through municipalities, departments and regions of Europe in 1999. Some hazards can also affect several states: flooding of border rivers (e.g. Rhine). The territory affected by the crisis or disaster resulting from the manifestation of the hazard is sometimes different from the territory of the hazard; such a situation is frequent in the case of an earthquake because of the nature of the rocks. For example, in the 1987 Mexico City earthquake, the effects were not concentrated in the area of the epicenter (defined as the projection of the focus on the Earth's surface, the point where the rupture occurs), as might be expected. The earthquake caused thousands of deaths and extensive damage in Mexico City, even though it was 400 km from the epicenter. The reasons for this are related to the nature of the subsoil; the loose lacustrine deposits of the Mexico City basin accentuated the effects of the tremors. The consequences of a dangerous event are also unequal according to the vulnerability of the affected territory, its accessibility, the density of the population concerned, its knowledge of the danger, and the types of construction, which themselves reflect social inequalities.

Origin of disasters and countries concerned	Date	Damage in billions of USD
Earthquake and tsunami, Japan	2011	210
Hurricane Harvey, United States	2017	180
Hurricane Katrina, United States	2005	125
Earthquake, Japan	1995	100
Earthquake, China	2008	85
Hurricane Sandy, United States/Canada/Caribbean	2012	68,5
Earthquake, United States	1994	44
Flooding, Thailand	2011	43
Hurricane Ike, United States/Caribbean	2008	38
Earthquake and tsunami, Chile	2010	30

Table 1.2. *Damage severity scale. The 10 most costly natural disasters since 1980 worldwide (source: assurland.com, 2017)*

The course of disasters related to natural hazards is generally circumscribed in space and time, although the consequences of such an event sometimes have longer effects. These are hazards of lithospheric origin: earthquakes, land movements and volcanic eruptions or hydroclimatic hazards: cyclones, storms, very heavy rainfall, heavy snowfall, hail or long-term drought. Heavy rains or snowmelt are sometimes responsible for flooding. Thus, in France, between 2001 and 2015, more than 1,300 natural events caused damage both in metropolitan France and in the overseas departments, resulting in more than 25,000 victims. Their cost amounted to about 27 billion euros. The 2003 heat wave represented a very deadly episode (about 15,000 victims), and the storms Klaus (in 2009) and Xynthia (in 2010) had a respective cost of 4.7 and 3.6 billion euros (see: https://www.catnat.net/).

These hazards are not distributed anywhere on the planet, they respond to geological, tectonic and climatic logics. Seismic and volcanic hazards are numerous on the Pacific border (Japan, Kamchatka, Alaska, California, Andes), at the contact of tectonic plates, earthquakes are frequent along the Mediterranean border. They affect Morocco (earthquakes of 1755, 1909, 1960), Egypt (1992) Portugal (1755), Greece (1810, 1928, 1953, 1978), Algeria (1716, 1980, 2003, 2010, etc.), Turkey (1939, 1983, 2011, etc.), France is not totally spared (1564, 1887 in Nice), then beyond, they appear in Iran, India and China. With the exception of California, Alaska, southern Europe and Japan, the areas affected by earthquakes and volcanism are in many developing countries where, when these processes occur, their effects are multiplied by the weakness of the infrastructure and an often insufficient preventive policy. Many cities have been or could be subject to disasters of lithospheric origin: Lima, San Francisco, Los Angeles, Mexico City, Tokyo, and many others, such as Algiers or Istanbul in the Mediterranean basin. In 2016, earthquakes in Taiwan, Japan, Ecuador, Italy and New Zealand caused USD 40 billion in economic losses, with only USD 7.41 billion in insured losses. The earthquake that ravaged central Italy in 2016 caused USD 5.6 billion in damages, with the losses borne by insurers only amounting to USD 320 million. In Italy, a country highly exposed to this type of hazards, only 1% of homes are insured against earthquakes. The Kumamoto earthquake (Ryukyu, Japan) in 2016 was responsible for nearly USD 20 billion in economic damages for which insurers only covered USD 5 billion.

At the forefront of climatic disasters are the consequences of cyclones that affect the oceanic facades located to the east of continents at tropical latitudes (Florida, West Indies, Bangladesh, Japan, Southeast China, Philippines, etc.). The violence of the winds and the heavy rainfalls which characterize them can cause victims and damages.

Cyclones are also very dangerous for populations when they occur in low-lying coastal areas. They can cause sea surges that can reach several meters, allowing the sea to penetrate far into low-lying areas and increase the effect of flooding often generated by cyclonic rains, as rivers have difficulty reaching the sea. Storms are frequent on European coasts (Xyntia in 2010 on the French coast). Heavy snowfall can occur in North America and block activities in Montreal, New York, Chicago. Heat waves also occur in temperate areas. Avalanches occur in most mountains as soon as the snowfall feeds a sufficiently thick and sometimes unstable snowpacks. The probability of occurrence of the phenomenon increases greatly with the occupation and use of the mountain, especially by skiers, who can cause snowpacks to break up. Disasters due to floods are largely the result of the settlement of populations in flood-prone areas (cities in developing countries, developed areas of the Indo-Gangetic and Chinese deltas, etc., and establishing of cities and economic activities in the valleys of industrialized countries).

All of these crises have significant costs, in terms of victims and damage to buildings, facilities, heritage, critical infrastructures and so on. The activities of the corresponding places must sometimes cease for fairly long periods, particularly when it comes to tourist activities. These effects are all the more important and damaging as the territory concerned is more vulnerable. A number of vulnerability factors can be seen in the example of Saint Martin (West Indies). Thus, the Caribbean island of Saint Martin affected by the passage of Hurricane Irma in 2017 reflects the consequences of the vulnerability of this space and its population, vulnerability prior to the course of Irma. In Saint Martin, if the physical data contributed to explaining the passage of the cyclone, the disaster that resulted is due to the fragility of people, facilities and the economy of the island. One of the main factors of vulnerability is the density of the population and the modes of occupation of the island, whose area in the French part does not exceed 53 km^2. The strong growth of the population from 8,000 inhabitants in 1982 to 28,500 in 1990 and 35,600 in 2016 is linked in particular to the development of tourism on the island at the end of the 20th century, itself related to the enactment of the Pons law in 1986 on tax exemption, which allowed people to benefit from a tax reduction on rental property investments made in overseas departments.

This development is taking place on the necessarily very attractive coastlines. However, the vulnerability of the population is all the more greater because a large part of the population of immigrant origin is impoverished and lives in informal, poorly built neighborhoods, where the fragile buildings are subject to the risks of submergence, flooding and violent winds. In 2010, 12,394 foreigners were living in Saint Martin, which represents about one-third of the total population. Most of these foreigners came from the Caribbean (77%), particularly Haiti (47%), Dominica

(16%) and the Dominican Republic (6%), and left their countries after natural disasters and because of poverty. A significant portion of the immigrant population has no professional training, sufficient knowledge of risks or means to protect itself effectively when a crisis occurs. The topographical characteristics of the island, whose very uneven central part, not very conducive to construction, is difficult to access, have historically pushed the population to settle in the immediate vicinity of the coastline, even "waterfront" that is sought by the actors of tourism and the population. Groups of dwellings or economic activities are established directly on the dunes. As these areas are reaching saturation point, urbanization is progressing on the heights. Saint Martin is now seeing an increase in construction on the hills (mornes), sometimes in the immediate vicinity of gullies which are frequently filled in during urbanization work. The edges of wetlands (ponds) have also been largely occupied by uncontrolled development, which has reduced their capacity to absorb water from rainfall, already reduced by siltation phenomena (Veyret 2017).

The anarchy of construction in Saint Martin is a major factor of vulnerability. The only urban planning document in force is a very old Land Use Plan, called the *Plan d'Occupation des Sols* (POS), which is little or not respected, including by the public authorities. A draft, called the Local Urban Plan (*Plan Local d'Urbanisme*, PLU), now exists, but what about the "de facto" constructions? Will illegal occupations cease with this new document? The mission conducted by the National Assembly (2014 mission report) emphasizes that the land registry is "in a dramatic state" and that the number of building permits filed in Saint Martin is ridiculously low (less than 200 per year), even though construction has been very considerable over the past few decades. According to the conclusions of the National Assembly report, it "would not be excessive to consider that more than half of the built heritage on Saint Martin has been built in an irregular manner and in particularly dangerous areas", as evidenced by the developments on the east coast of the island where many hotels and summer rentals are located. On the west coast, hotels and restaurants have been built directly on the beach, and the area known as "Sandy Grounds" located southwest of Marigot, consisting mainly of hard-wall housing built in an uncontrolled manner, is located less than one meter above sea level, on a thin strip of land between the coast and the large lagoon of Saint Martin. For the population, and many elected officials, it is possible to build anywhere, including in the "50 geometric steps" zone. Consisting of a strip of land 81.20 meters wide determined from the limit of the shoreline, the "50 geometric steps", created by Colbert, define a coastal strip. It was used for military purposes to protect the inhabitants, and belonged to the state public domain. Over time, parcels of land in the 50 steps zone were appropriated, particularly after the promulgation of law No. 55-349 of April 2, 1955, which placed this zone in the state private domain, making it alienable and prescriptible. The littoral law No. 86-2 of January 3, 1986 again classified the "50 steps" zone in the state maritime public domain, prohibiting any transfer and any private appropriation. However, the land located in this area and belonging to private or public persons who can prove their title of ownership are considered as not being part of the state maritime public domain. In addition, law No. 96-1241 of December 30, 1996 relating to the development, protection and enhancement of the "50 geometric steps" zone in the overseas departments has led to the regularization of the situation of untitled occupants, by allowing them to acquire the land they occupy for their main residence, in order to improve the hygiene of buildings often built in precarious conditions. However, if we follow the work of the ONERC (*Observatoire national sur les effets du changement climatique*, National Observatory on the Effects of Climate Change), the consequences of climate change on coastal areas should result in particular in the acceleration of marine submersion and coastal erosion, the strengthening of the frequency and intensity of storms, and changes in the wave regime and ocean circulation.

The vulnerability of the island is largely related to political and social factors. The French part of Saint Martin was established as an Overseas Collectivity (*Collectivité d'Outre-Mer*, COM) under Article 74 of the Constitution, by organic law No. 2007-223 (February 21, 2007), amended in 2010 (organic law No. 2010-92). Saint Martin was previously administratively attached to Guadeloupe as a simple commune, but now has its own administrative organization, powers and legislative regime. The community of Saint Martin has a territorial council elected by universal suffrage, a president assisted by an executive council. The community has an economic, social and cultural council. A delegated prefect, representing the French State, is responsible for national interests, compliance with laws and treaties, public order and administrative control. Saint Martin has a magistrates' court and a deputy public prosecutor. The community exercises and combines the powers previously vested in the commune, the department and the region of Guadeloupe. It is also the beneficiary of state competences, notably concerning taxes, duties and fees, road traffic and road transport, property rights, access to work for foreigners, tourism, urban planning, construction, housing and energy. It does not have jurisdiction over the environment. The role of the state in environmental matters is to ensure that laws are enforced and that public policies are implemented in the relevant areas. However, the community of Saint Martin has the possibility of adapting the laws and regulations in force in environmental matters to the island's specificities. In 2017, the highly indebted COM had hardly implemented the regulations on construction and risk. The use of the territory for the sole purpose of developing waterfront tourism for the exclusive benefit of a few, largely explains the consequences of the course of Irma. The balance of destruction and damage amounted to 3 billion euros, including 1.9 billion insured losses (Irma CCR 20178), making this event one of the most expensive for the French insurance system.

Overall, 95% of buildings and infrastructure were damaged. The prefecture, the multimedia library and four of the 21 schools were considered too dangerous to be occupied again (Moatty et al. 2019). The intertwining of physical, human and economic factors must be considered in its entirety. The disaster expresses the physical event and its dramatic consequences and justifies the implementation of an adapted management that is part of a specific social, economic, institutional situation indicative of crisis and high vulnerabilities. Thus, from a brief and well-localized event (a cyclone or an earthquake, for example), a disaster occurs, which is part of a more or less vulnerable context at the origin of a crisis of variable duration and magnitude.

1.3. Localized crises of natural origin aggravated or provoked by human activities

Other crises may result from hazards whose effectiveness is increased by human activities and development. This is the case of accelerated soil erosion, which has been well described in Africa (Roose et al. 2012), South America and Europe, as well as the complex case of desertification, which is specific to semi-arid regions and which Mainguet (1995) emphasizes "is the result of natural mechanisms that are exacerbated or induced by man; it manifests itself by a deterioration of vegetation and soils and leads, on a human time scale, to an irreversible reduction or destruction of biological potential". The notion of irreversibility is limited to the scale of a few generations, because beyond that, other methods of tilling the soil could allow its use again. It is fundamental, since it prohibits classifying among desertified areas or areas in the process of desertification all those that simply record an acceleration of soil erosion. Desertification sanctions the rupture between the environment's resources and the demand of societies. These ruptures are frequent and repeated, especially in semi-arid regions normally characterized by a relatively limited biomass. The sedentarization of populations, overgrazing, the high demand for wood and poorly managed resources lead to the formation of increasingly large haloes of plant cover and soil degradation around cities and towns, and the wind can then mobilize fine soil elements to build dunes. In the Sahel, the impact of droughts, although recurrent, has been reduced or even non-existent as long as anthropic actions have remained small in scale. Thus, according to Le Houérou (1993), "the state ranches found in most Sahelian countries and which are subject to moderate loads, have undergone nearly 20 years of almost continuous drought since 1969 without serious prejudice". Droughts alone cannot be exclusive factors of desertification; meteorology has played the role of revealing the imbalance between the rapidly growing population, its sedentarization, at least in some cases, and the available resources. The degradation of vegetation cover, an indicator of desertification in the Sahel, contributes to increased drought by accelerating desiccation and soil destructuration, in other words, accelerating the crisis.

Among the risks aggravated or caused by human activities, wildfires, especially in Mediterranean areas, are a threat to the environment and to populations, especially those who have more or less recently moved into the forest environment, as Clément (2005) points out.

They are often neo-rural, temporarily or permanently installed (second homes, retirees, urban dwellers), without any real links with the forest environment. These new inhabitants of the forest have no memory of the risk. The forest is a simple decoration for them. It symbolizes a rediscovered freedom from the constraint of the city, in the same way as the rivers, whose proximity is sought after despite the risk of flooding (De Vanssay 2003). The progression of urbanization around the large Mediterranean cities and the sprawl of the forest by villas in the coastal tourist areas have not created the problem of fire. On the contrary, they have certainly opened up a new fire front by multiplying the forest-urbanization interfaces (Clément 2005). Since the end of the 20th century, the increase in the number of fires around the Mediterranean has also resulted from the increase in the number of people visiting the forest and activities in the forest or on its margins. The structure of forests stands (age, species, type of ownership, etc.) and their maintenance must also be taken into account, along with seasonal climatic events, that is, episodes of drought and extreme heat, when analyzing forest fires. It is difficult to estimate the direct cost of these fires. As an example, the journal "Mediterranean Forest" considers the cost of the 2003 fires in Portugal to be more than 1 billion euros for 430,000 hectares of forests burned (Vogel 2019).

1.4. Industrial, technological and nuclear disasters and crises: localized crises and spatial consequences

The term technological crisis is used to consider the negative consequences of the use of technologies born in the industrial age. These can be crises related to the storage, production and transport of energy and hazardous materials, dam failures (Malpasset France 1959, Langarone Italy 1963, Banqiao China 1975, Val de Stava Italy 1985, Brumadinho Brazil 2019) and accidents to ships (Amoco Cadiz, etc.). Industrial crises are associated with malfunctions of factory activities. The potentially dangerous establishments are in many cases inserted in the urban fabric, which makes us fear chain effects, a series of events difficult to predict and control. Industrial and technological disasters are generally unexpected, sudden, and sometimes very damaging because of their human, social, economic and ecological consequences. These disasters result either from an explosion, the shock wave causing an overpressure wave whose effects can affect people, vegetation and buildings, or from the leakage of a toxic product that causes various types of pollution, or from a fire that causes burns and pollution from the smoke and the products released. The effects can be seen around the point of impact, but often affect larger areas. These crises are rare, and each has its own specificities linked to the activities themselves. The dramatic explosion of the Feyzin refinery in 1966, responsible for 18 deaths, led to the discovery of the BLEVE (Boiling Liquid Expanding Vapor Explosion) phenomenon, which is now taken into account by the industrialists concerned and by the safety and technological firefighting teams. The AZF explosion in Toulouse (31 dead, 2,500 injured) had dramatic effects on property and nearby buildings. Nearly 1,300 companies were damaged, representing 20,000 jobs, as well as homes, schools and so on. As a result of these crises, new practices and regulations were introduced (risk reduction at source, hazard studies, etc.). A follow-up of industrial and technological accidents is possible thanks to the ARIA database (*Analyse Recherche et Information sur les Accidents*), which keeps information on accidents and major pollution that have occurred in facilities and that may affect the environment, safety or public health. In France, plans for the prevention of technological risks (*plans de prévention des risques technologiques*, PPRT) have been created. The promulgation of the law of July 30, 2003 on the prevention of technological and natural risks and on the repair of damage, a response by the legislator to the AZF disaster led to the development of such plans (PPRT) concerning the 670 so-called "high threshold" Seveso sites (named after European directives enacted following the dispersion of a toxic cloud of dioxin in 1976 over the Italian city of Seveso). The PPRT for the Vallée de la Chimie in Lyon, which was approved in 2016, defines areas with high constraints for the poorest neighboring populations in the Lyon metropolitan area. This plan regulates future and existing urbanization around the 10 Seveso-classified high threshold establishments in the Chemical Valley, in the municipalities of Lyon 7e, Pierre-Bénite, Saint-Fons, Irigny, Feyzin, Vénissieux, Oullins, Solaize, Saint-Symphorien-d'Ozon and Vernaison. Initiated in 2009, it is the result of close collaboration between the industrialists who are the source of the risk, local authorities, government departments and the public. It establishes safety standards for 26,000 residents and 12,000 employees exposed to the effects of a possible industrial accident. The objective of the PPRTs is to resolve inherited situations: the proximity of factories to urban areas, since most industrial sites, which were initially located on the outskirts of towns, have gradually been integrated into the urban fabric as towns have grown; residential neighborhoods, which were often built well after the factories, are located next to them. The objectives of these plans are also to limit urbanization around high-risk facilities, to keep homes away from them and plan work on existing homes that are highly exposed. Thus, the PPRT prohibits or limits the construction of infrastructures, extensions or new buildings in certain areas close to potentially dangerous facilities. In these areas, expropriation procedures may be applied or the right of relinquishment may be exercised, allowing owners to sell their property to the municipality. In more remote areas, the PPRT requires buildings to be reinforced to withstand thermal or overpressure effects: installation of triple glazing, construction of a containment room, reinforcement of roofs and so on. All of these measures must be financed by the state, local authorities, the manufacturer and local residents. However, the methods of application and financing of the PPRT are sometimes contested before they are applied, as is the case at La Mède near Fos-sur-Mer, around the Total plant. How can we explain this opposition, whose objective is to reduce risks and protect the population? One explanation is the refusal of local residents to contribute financially to the protection of their homes, since the law requires that safety work be carried out on homes located in high-risk areas if, after applying the best available "economically acceptable" practices and techniques, the danger zones have not been sufficiently reduced. They consider that this is the responsibility of the manufacturer. At La Mède, the population, composed in part of retirees from the plant, is attached to it, and the local residents, who are not opposed to maintaining the industrial activity, are not asking for the relocation of the plant and so on. In the areas defined by the PPRT zoning, the territorial identity and real estate heritage are devalued because of exposure to risk. The evolution of the peri-industrial districts, as foreseen by the PPRT, should lead to the departure of at least part of the population, those who manage to sell their property. The renewal of the population favors the installation of a socially disadvantaged population, attracted by a low real estate market. This evolution leads to a double inequality: environmental inequality defined as an inequality of exposure to pollutants, nuisances and environmental risks, and social inequality, with only the poorest remaining in the dangerous areas inhabited by working class or marginalized populations. When the accident occurs, the crisis takes on a social dimension by mainly threatening these disadvantaged populations, who, in part, worked within the plant and therefore lose their activity (Provotolo et al. 2015).

With regard to the transport of dangerous products at sea, the disaster caused by the sinking of Amoco Cadiz on March 16, 1978, following a breakdown, demonstrates the extent of the consequences for an entire region and for the maritime and coastal areas affected by pollution. The Liberian oil tanker Amoco Cadiz ran aground on the rocks of Portsall, in North Finistère, loaded with 223,000 tonnes of crude oil and 4,000 tonnes of bunker fuel. The entire cargo escaped, polluting 360 km of coastline between Brest and Saint Brieuc. This disaster led the government to overhaul its response plan (the Polmar Plan), and impose traffic regulations in the Channel. After this disaster, the *Centre de Documentation, de Recherche et d'Expérimentations* (*Cedre*) on accidental water pollution was created.

This research center worked on the types of pollutants and the steps to take to limit their impact on the environment. The consequences of the disaster went beyond the vast polluted coastal and maritime area to reach activities and people. The ecological and economic losses were considerable: costs of clean-up operations, costs of restoration trials of coastal environments and ecosystems, economic losses suffered by fishermen (1,300 concerned) and oyster farmers; losses of amenity suffered by consumers, local residents or visitors (tourism) of the areas affected by pollution. The state and the affected municipalities engaged in a long and difficult lawsuit in the United States against the Amoco company. After 14 years of proceedings, they finally obtained 1,257 million francs in compensation, that is, a little less than half of the amount requested.

Date	Location	Type of accident	Victims and damages
1973	Saint-Armand-Les-Eaux (North)	Rollover of a semi-trailer transporting propane. Formation of a propane cloud, fire and explosion of the tank	nine dead, 45 injured. nine vehicles and 13 houses destroyed. Scattering of debris in a radius of 450 m
1978	Los Alfaques (Spain)	Explosion of a semi-trailer of propylene on a road along a campsite	216 deaths
1997	Port-Sainte-Foy (Dordogne)	Collision at a level crossing between a tanker truck carrying 31 tons of petroleum products and a railcar. The fire spread from the tanker to the passenger cars and a house	12 dead, 43 injured

Table 1.3. *Some 20th century disasters related to land transport (source: georisques.gouv.fr)*

Crises of nuclear origin can have a triple origin, they can result from accidents occurring in power plants, from waste management or from problems related to military nuclear power. Three Mile Island in the United States in 1979, Chernobyl in the Soviet Union in 1986, Fukushima in Japan in 2011 are the emblematic of crises or disasters that occurred in these three plants. The accident at Three Mile Island was the result of a core meltdown in one of the power plant's reactors, a situation that was considered unlikely at the time, and which made people aware of the dangers associated with the operation of nuclear reactors. In Japan, in Fukushima, following the tsunami caused by the 2011 earthquake, the emergency cooling systems of the nuclear reactors stopped, which caused the partial meltdown of the cores of three reactors and significant radioactive releases. This nuclear accident was classified at level 7 of the INES scale (International Nuclear Event

Scale; the events of levels 1–3, which are without significant consequences on the populations and the environment, are qualified as incidents, those of higher levels (4–6), of accidents and serious accidents, while the disaster of the nuclear power plant of Chernobyl on April 26, 1986, as well as that of Fukushima are major accidents of category 7). In 2015, the Japanese government reported 118,812 displaced persons, but this number seems to have been minimized, as a certain number of people did not declare their displacement for fear of losing certain rights, in particular free medical care. Protocols have been put in place in order to assess the health status of the population after the disaster; however, doubts remain about the reliability of these follow-ups and their conclusions.

From the point of impact of the accident (technological, industrial, nuclear), which is fairly easy to define, the effects of the associated disasters can extend far beyond the initial area, reaching, in the case of the Chernobyl nuclear pollution, Western Europe and beyond, affecting the quality of the air, soil and water, in other words, affecting plant formations and populations (including through food and polluted water). In all cases, these crises are characterized by the importance of the uncertainties generated by the rarity and scale of this type of event, which makes it difficult to estimate the probabilities of occurrence and the damage.

1.5. More diffuse environmental, food and health crises on a regional or continental scale

It was not until the development of the hygienic movements in the 19th century that the relationship between environmental quality, particularly the quality of water, and epidemics became apparent. The great cholera epidemic that affected the Ruhr in 1910 made it possible to establish the consequences of untreated urban water discharges into low-flowing rivers and led to the creation of river cleanup syndicates in this region.

Various nuisances can also cause or reinforce health crises. Thus, the question of waste management is posed in olfactory, visual and hygienic terms. The relationship between waste and health is certainly not recent, since these concerns for public hygiene and healthiness have justified restrictive legislation concerning the public disposal of waste and the progressive organization of a public management service (Bertolini 1990). In addition to collection, there is the question of waste treatment, which is inappropriate and causes health crises (release of dioxin from incinerators). Thus, in 2007–2008, the issue of waste in Naples and more broadly in Campania was the cause of "an environmental and economic crisis, but also moral and social". In late 2007, the city of Naples was in a catastrophic "emergency waste" situation, inducing bad odors, proliferation of animals (rats, flies, etc.) and possible development of diseases: cholera, salmonella, leptospirosis and so on. The situation that had lasted for over a decade took on an unprecedented magnitude in 2007, threatening the health of populations. The foreign media took hold of this situation, which was widely broadcast, and had consequences on the regional economy through tourism, which was reluctant to "brave" the nuisances linked to the waste. Politicians have been blamed for the delay in planning, the inadequacy or absence of suitable landfills, inadequate treatment of urban waste when it exists; but it is especially the involvement of Camorra, the local mafia network, which given the interests at stake, has largely appropriated the waste to the detriment of an acceptable management of the environment, but for very considerable profits (Giuliani 2009; Larroque 2014).

The World Health Organization (WHO) associates environmental quality and health, although quantifying the links between the physical, biological and social environment and health is difficult to envisage. It is sometimes difficult to distinguish between endogenous and exogenous factors of diseases and define the exact part that belongs to the environment. Setting environmental quality standards to reduce or eliminate risk involves establishing indicators. For this purpose, in the case of air pollution, it is necessary to model dose-response relationships, early markers of effects and chronic effects. Pollution of various kinds has been a source of disease in the past; air pollution by sulfur released from district heating during four days in December 1952 in London caused the death of several thousand people. A distinction is made between primary and secondary pollutants. Primary pollutants come directly from pollution sources (road traffic, industry, heating, agriculture, etc.): carbon oxides, sulfur oxides, nitrogen oxides, light hydrocarbons, volatile organic compounds (VOCs), particulate matter (PM10 and PM2.5) and heavy metals (lead, mercury, cadmium, etc.). Secondary pollutants are the result of chemical reactions between gases, such as ozone and nitrogen dioxide. Thus, ozone associated with other photochemical pollutants (PAN or peroxyacetyl nitrates, aldehydes, etc.) is composed of a brownish cloud, which sometimes stagnates over large cities and is responsible for respiratory system diseases (bronchiolitis). This pollution raises the question of tomorrow's urban organization and traffic within cities. Many European cities, such as Athens, Milan and Paris, limit the speed of cars in the event of pollution peaks, and in some cases reduce traffic in the city center by means of urban tolls (London). Water pollution is responsible for many diseases: dysentery, cholera, tropical diseases related to water (malaria, schistosomiasis, etc.). Health crises are expressed in different ways, especially in developing countries, such as malaria and HIV-AIDS. These crises, which are often poorly treated, reflect social, economic and environmental shortcomings, incomplete access to education and hygiene, and even insufficient or deficient nutrition. According to the WHO, 45% of the deaths of children under the age of five are due to nutritional deficiencies or malnutrition. In addition to this, there are often insufficient health services and a low density of doctors in many developing countries. In its 2018 report "A universal truth: no health without human resources", the WHO underlined the lack of 7.2 million health workers worldwide, a figure that is expected to reach 12.9 billion in 2035. Health statistics from 2016 show that many countries are still far from universal health coverage, which allows everyone to have access to health services without incurring financial hardship. Yet, we know that every year, 100 million people fall into extreme poverty (USD 1.9 per day or less) due to health care costs.

Food crises are similar to health crises; they are of several types, some resulting from the presence of pollutants in food products: food additives, pesticide residues and fertilizers that can affect the entire food chain. These risks refer to the notion of consumer exposure, which is often poorly understood. Exposure is cumulative but must also take into account the variability of food composition: variability of heavy metal content depending on the fishing grounds, for example. The variability of consumption of each food is also involved. It is necessary to set tolerable daily doses, toxicological limits. Individual variability is a delicate problem and shows that thresholds are difficult to apply to all. Large-scale food crises mainly concern developing countries, but they can also affect the poor in rich countries. These crises correspond to geographical areas where undernutrition or malnutrition is rife. The great famine-type crises that have marked the history of societies are now a thing of the past. In the 20th century, there were still major famines linked to political factors and armed conflicts (Russia 1921–1922, China 1959–1961, Biafra in the 1960s), all of which caused millions of victims. However, undernourishment persists and still concerns nearly one billion people. This is what we call "visible hunger", affecting people who do not have enough food to reach the 2,000 to 2,500 calories per day per person. This "visible hunger" must be distinguished from "invisible hunger" corresponding to a diet deficient in vitamins and minerals. Such situations are primarily due to poverty and the distribution of resources, rather than to a shortage of them. The current agricultural production, if it were more equitably distributed, would make it possible to correctly satisfy the needs of all. The existence of food is not enough, it is also necessary to have access to it, which is not always possible if politicians choose to sell these resources on the world market. The increase in food prices on the world markets in 2008, linked to the financial policies of large corporations, triggered a "food crisis", particularly in Africa where some countries are highly dependent on imports (especially rice). Some food crises are not the result of a lack of food, but of speculation that leads to the non-distribution of food so that prices rise. In 2008, for example, the price of Thai rice on the world market reached 1,000 USD per ton, three times the price at which it was sold six months earlier. This spectacular increase due to speculation provoked demonstrations by the poorest city dwellers in the capitals of certain countries of the South. These were referred to as "food riots". The price of rice is not the only one at fault, since the price of other basic cereals (wheat, corn) and oilseeds (soybeans, palm oil) has undergone the same evolution after the surge in oil prices. Hunger and the associated crisis is first and foremost a social and economic issue before being linked to agricultural production and nature, including in countries where agriculture is still closely subject to weather variability. Environmental factors, such as drought can worsen the living conditions of societies and increase risks. Nevertheless, a disaster occurs when "the rigidity of the socio-economic system does not allow the necessary adjustments or when socio-political disorders do not allow any adjustment". Malnutrition is also responsible for health crises whose magnitude is increasing; it is obesity that has increased between 1975 and 2016 at an accelerated pace. Obesity, which is linked to an industrial diet that is too rich and unbalanced (in sugar, fat, salt), reduces life expectancy. It is primarily a disease of the poor that is particularly prevalent in urban areas of middle- and high-income countries (Lançon et al. 2008).

1.6. "Reticular" crises

Some crises are difficult to locate precisely; they affect economic, political and social sectors and require a reticular analysis.

The terrorist crisis, as defined in the International Convention against Terrorism (1937), which was later supplemented and clarified, is "a criminal act directed against a state or intended to cause terror among specific individuals, groups of individuals or the public". The U.S. State Department defines terrorism as "premeditated, politically motivated violence perpetrated against non-combatant targets by subnational groups or clandestine agents in order to influence a public". For Blin (2005), "a terrorist act is a political act whose purpose is to destabilize a government or political apparatus, where the psychological effects sought are inversely proportional to the physical means employed and whose primary, but not exclusive, target is the civilian population". Terrorist groups seek to reach a security-conscious public, and they do so within a democratic framework where the citizen "is as important a figure in the state as the ruler". The driving force behind terrorism can be ideological, social, cultural or religious. There are more than a hundred definitions of terrorism, but a few general features can be highlighted: terrorism is a form of violence, its nature is specific and its forms are themselves multiple. The crisis of terrorism, which is the cause of a state of stress for the population undergoing it, directly affects the rule of law, and is linked to networks deployed in a large number of countries, and widely relayed by social networks. Terrorism is based on a reticular organization, but acts on spaces that are particularly chosen for their vulnerability (auditoriums with large numbers of people) or the symbols they represent (*World Trade Center*). Terrorism requires a maximum relay by the media and social networks. Terrorism is not a new risk: more than 10,000 people have lost their lives and more than 60,000 have been injured as a result of terrorist acts in the last 30 years worldwide (Michel-Kerjan 2001; Deffarges 2003). However, the attacks of September 11, 2001 in the United States, which left a particular mark on the world, affected the entire international community. They concerned a state that had been largely sheltered from terrorism; they took place in a peaceful country and were broadcast live by the world's media. They claimed many lives, more than 3,000 in the collapse of the two *World Trade Center* towers, the crash of planes into the Pentagon and Pennsylvania and among the emergency services. The insured losses amounted to USD 40 billion, making these attacks the most expensive loss in the history of global insurance and reinsurance (Swiss Re 2002). These events resulted in a 150% increase in the US defense budget between 2001 and 2011 (USD 750 billion in 2011). Terrorist attacks can involve information systems and critical infrastructure. After the terrorist attacks of September 11, 2001, the attacks in Madrid in 2004 and London in 2005, and those in Paris in 2015, the protection of critical infrastructures has become a priority for different governments or organizations around the world.

Crises related to critical infrastructures (telecommunication lines, water, energy, transportation, etc.) will also be highlighted. Bouchon (2011) defines "critical infrastructure as physical facilities and [information] technologies, networks, services and assets that, if disrupted or destroyed, can have serious impacts on the health, safety or economic well-being of citizens or the work of governments". The same author distinguishes three categories of critical infrastructure: those that contribute to the continuity and credibility of government activity and the maintenance of fundamental strategic services: defense, health, administration, police and emergency services, etc.; those that contribute to the maintenance of order and the provision of basic public services: water supply, energy, food, agri-food and industrial activities, education; and finally, those that are the responsibility of the private sector in general and also provide basic services (transportation, telecommunications and information, banking and financial services, insurance, security of private services). "Territorial criticality" expresses the fact that critical infrastructures are not critical in themselves, but in relation to a socio-economic, political and geographic context. They belong to reticular territories of "vulnerability basins" to be considered at different scales (European and national). From a geographical point of view, critical infrastructures correspond to complex systems that can be analyzed in terms of networks and associated reticular territories, characterized by the connection of discontinuous spaces. The systemic approach therefore responds to the need to take into account the different actors, the scale and the temporal context of analysis. Characterizing the territories specific to critical infrastructures allows us to highlight their relationship with the administrative territories of crisis management. The degree of criticality of a system makes it possible to consider the vulnerability of a territory, which is based on the degree of direct exposure of a critical element, combining its intrinsic vulnerability linked to its size, its specificity and so on, and its vulnerability to internal or external hazards. Networks do not create homogeneous territories: they value and prioritize discontinuous places that do not all have the same accessibility to the critical service.

System analysis	Systemic approach	Spatial and cartographic approach
System function	Nature/quantity/quality of service provided	Area served by the service (according to the scales)
Composition	Subsystems and basic components forming the system (technical, human, organizational, etc.)	Location of the support infrastructure, distance between components
Organization	Communication channels and hierarchy of relationships between components (connectivity, connectedness)	Types of spatial organization of the network (graphs, density)
Process	Types of flows, variations over time	Orientation, direction of flows, places of origin, and arrival and variations over time
Relationship with the environment	Types of relationships with the environment	Location of administrative boundaries, geographical context, economic, etc.
Interdependencies	Types of interdependencies (logical, cyber, physical, geographic)	Location of interdependencies

Table 1.4. *Systems approach and mapping approach*

Critical infrastructures are essential to the functioning of society, yet their disruption and/or failure are likely to jeopardize the security of individuals, and the economy of a region, or even a country (Galland 2010; Lhomme 2012; Laganier 2015). It is therefore necessary to protect these infrastructures in order to first prevent and then mitigate the crisis that would be triggered by their reduction or discontinuation of their operation, which is essential to safeguard the vital interests or essential needs of all or part of the country or population. For example, a blackout that occurred in North America in August 2003 deprived about 50 million users in several American States and the province of Ontario in Canada of power for four days and so on. The blackout occurred in the New York subway system and affected industrial facilities in the affected areas. It cost between four and 10 billion US$. The earthquake of March 11, 2011 followed by the tsunami that hit eastern Japan also affected critical infrastructure, particularly in the energy sector: six refineries out of 28, 29 fuel depots out of 190 and 680 gas stations ceased operations. A total of four million people were deprived of electricity on 11 and 12 March, leading to the total or partial shutdown of several industrial facilities: cement factories, refineries, drinking water pumping stations, chemicals and food processing. Over the following months, more than 10 million Japanese users were affected by the load shedding operations. As 19 drinking water pumping stations were partially or totally destroyed, 2.23 million households were deprived of drinking water in the first days after the disaster.

Since the end of the 20th century, governments have identified critical infrastructures and ranked their vulnerability by underlining the interrelationships that can exist between them. Among these, the geographical proximity of two CIs (Critical Infrastructures) can, in the event of a problem at one of them, affect the other (e.g. explosion of an industrial site next to a railroad); the dependence of one CI on another is another aspect (e.g. gas supply for electricity production). A major flood of the Seine in Paris would have consequences on critical infrastructures, metro, railroads, drinking water treatment and distribution, wastewater treatment, electricity distribution, waste collection and treatment, and health problems (hospitals and institutions for the elderly in flood-prone areas). The OECD report published in 2014 considers that 1.5 million users would be deprived of electricity, 140 kilometers of subway lines would be closed preventively, five million subscribers could suffer prolonged water cuts and 1.3 million a deterioration in its quality. The induced effects on the entire French economy are evident. The costs of direct damage are estimated at between three and 30 billion euros.

1.7. The “mega-crises” on a global scale, the domino effects

These are “out-of-frame” crises (Lagadec 2012) or mega-crises representing a new type of crisis, which concerns the whole of society and has repercussions in a large part or even the entire world. Until now, many crises, and in particular those of natural and technological origin, have been limited to a more or less vast space often localized near the place of impact, whereas in “mega-crises” we speak of a systemic impact (World Economic Forum 2014) with a global scope. According to Gérard Pardini, these “mega-crises” of the 21st century are the result of the existing interconnections between the different countries of the world; they are therefore closely linked to globalization, whose interconnections are amplifiers of crises, as evidenced by the 2020 pandemic. The origins of these mega-crises can be multiple. Thus, the earthquake in Japan and the associated nuclear accident, already mentioned, took on the dimension of a mega-crisis when the consequences of the event became apparent at the international level, and the cuts in the production chains of major industrial groups had consequences for the Asian economy and the United States. Japan's position in high-tech sectors (it supplies 73% of navigation systems for automobiles), explains why the shutdown of its production caused supply disruptions in the electronics and automotive sectors (Sampognaro et al. 2012).

The very causes of these major crises can be out of proportion to their scale. The sharing of and access to certain resources, whether renewable or not, generate crises that can result in latent or open conflicts, as is the case with oil resources. The decrease or fall in the price of certain resources and the economic consequences of such situations can lead to a deterioration of the social fabric, an increase in insecurity and the shaking of the foundations of society.

A mega economic and social crisis can have multiple causes, some of which correspond to factors that may seem minor. Thus, a crisis may result from the choice made by farmers to plant a particular crop or production channel, or to engage in processes of technical change. Depending on the sales channels, markets, local or broader exchange systems, the weight of subsidies that determine the price of agricultural products, the role of stock exchanges (Chicago, etc.) that set the price of certain products (cereals, etc.), and the consequences of simple agricultural changes can be translated in social and spatial terms that go far beyond the area where the changes took place and so on. The spatial consequences of such situations can go as far as the displacement of populations and permanent emigration.

Economic crisis can also affect companies, as a result of competition, national or global choices that are themselves dependent on sales or production networks. The consequences can be closures, with effects on employment and on the local economy. The opening up of economies to globalization, the deregulation of economic life that has taken hold since the early 1980s, and the emergence of Internet networks and social networks add to this evolution and result in the development of a virtual market. Space and time are in a way abolished, the financial sphere is unified by the language (English), the currency (the dollar), the method (maximization of short term profits) and the culture of Wall Street or Chicago. This justifies the possibility of large profits and also of enormous losses, leading to a systemic crisis resulting from a context of extreme complexity and also of extreme fragility, where a simple technical malfunction can have enormous financial consequences. The financial system thus evokes a sort of house of cards that can collapse from one moment to the next following a simple rumor, as witnessed by the financial crisis of 2007, initially limited to the United States, then exacerbated by the bankruptcy of the Lehman Brothers in 2008, with the "subprimes, the excesses of consumer credit and the existence of liquidity that was disproportionate to the real capacities of borrowers, as well as the 'securitization' of assets, some of which turned out to be 'rotten'" (Wievorka 2009). This crisis very quickly affected the entire global economy (Outin 2013). In France, GDP growth, which had been +2.3% in 2007, was only +0.1% in 2008 and became negative in 2009 (-2.5%). In the euro zone, 2009 was marked by an even greater decline in wealth produced (-4.1%). Greece, Ireland, Cyprus and Portugal received aid conditional on the implementation of structural reforms in the form of company privatizations, public employment cuts and labor market flexibility. These measures led to much discontent in these countries, and also had a rapid impact by bringing economic growth in Europe to a virtual standstill. According to INSEE, the unemployment rate reached 9.9% at the end of 2012 in Metropolitan France, compared to 7.2%. This financial crisis was followed by an economic crisis and social difficulties that affected all countries of the world. The crisis of 2007 is also part of a longer period of time; it would only be a moment in a transformation that began, in several countries, with the end of the 30 glorious years, from the mid-1970s and the oil shock of 1973. It was the end of the welfare state, of controlled finance, of the smallest possible inequalities (Wieviorka op.cit.).

The great pandemics also fall into the category of mega-crises. Without mentioning plagues, smallpox, pandemics still occurred during the 20th century, such as the Spanish flu, which came from Kansas in the United States and whose name came from the King of Spain, Alphonse XIII, raged from 1918 to 1919. It was due to a virus that transferred from animals to humans. This epidemic spread worldwide during 1918 in several waves, with a virus that mutated and became more virulent. According to the Pasteur Institute, it killed more than 30 million people.

Since the 1918–1919 flu, other flu pandemics occurred: in 1957–1958, the so-called Asian flu and in 1968–1969, the Hong Kong flu. The Asian flu is linked to a virus, originating from a mutation of wild ducks and a human strain of flu. It was first identified in China in 1956. Reaching Singapore, then Hong Kong and finally the United States within months, this flu caused between one and four million deaths worldwide. The strain then evolved into another pandemic from 1968 to 1969, with the Hong Kong flu killing approximately one million people. The Covid-19 pandemic of 2020–2021 shows the global implication of such a situation, not only in terms of the number of victims, but also in economic, social and political terms.

The "mad cow" crisis, which took place in two stages, in 1996 and then in 2000, is interesting to consider because of its media treatment and the unreasoned fears associated with it (Masson et al. 2003). The number of victims was reduced, but consumer reactions were varied, "due to the uncertainties and contradictions" of a scientific knowledge that was not sufficiently "stabilized" on the question of the transmissibility of the prion from cows to humans (Peretti-Watel 2001, 2004). Doubts exist at all levels of production and decision-making, from politicians to farmers to agri-food industry managers. This question of confidence refers to the cultural and political stakes of the crisis. Peretti-Watel underlined that "trust is unequally distributed within the population": the conspiracy thesis "seems to concern more vulnerable individuals, located at the bottom or at the periphery of the social structure" (Peretti-Watel op. cit.).

1.8. Conclusion

The notion of crisis covers multiple realities. Each crisis is unique. While the triggering event may have a variety of origins, it above all reveals dormant vulnerabilities, specific to the territories concerned: material, functional and structural vulnerabilities. A crisis also highlights the importance of the interrelationships that drive any territorial system, the possible negative feedback loops and other domino effects which, if not sufficiently addressed during the crisis prevention and preparation phase, can constitute aggravating factors during extreme events.

Although each crisis is unique, they all share a number of common characteristics. These include the disorganization of the normal functioning of the territorial system, which can go as far as chaos, and the difficulty in making decisions to deal with the unexpected in a context of significant uncertainty, or even lack of knowledge about the origin of the crisis and its effects. Crises are also marked by the fact that people and even decision-makers are stunned by the brutality of the events and the complexity of the crisis to be managed; the processes of preparation for the crisis and anticipation are then taken for granted. From then on, crises are not only disruptions; they are also moments of rupture in the organization and functioning of territorial systems, ruptures that can lead to a different way of thinking about crisis preparation and post-crisis reconstruction. These are the challenges of modern territorial crisis management (Lagadec 2012; Rey 2015).

2

General Principles of Crisis Management

Freddy VINET[1] and Anne-Marie LEVRAUT[2]

[1] *Université de Montpellier III, France*

[2] *Conseil Général de l'Environnement et du Développement Durable, France*

Crisis management preparedness commonly combines three elements: crisis planning, that is, the anticipated definition of an organization, exercises to train people and test that organization, and the anticipation of resources needed for backup, rescue and recovery missions (food supply for the population, shelter, supply of materials, etc.). These elements of preparation and procedures are used by crisis management professionals in country-specific arrangements (Corack et al. 2018). We will not describe these different components in detail here because we believe that a crisis occurs precisely when these procedures, resources and mechanisms are not adequate to stop the increase in damage (victims, material damage, etc.) (Lagadec 2006; Reghezza-Zitt 2019). We therefore question the meaning of the expression "crisis management" from the outset. The expression is convenient, but it is undoubtedly an oxymoron. Indeed, if we define a crisis as a rupture, as a moment when the usual procedures and means of managing the disturbances caused by an event (health, natural or technological) are exceeded, then we are no longer in management sensu stricto. Thus, it is necessary to question the context of crisis management. Crisis management is above all the search for the means to regain control of a situation or to avoid losing it. For Staupe-Delgado and Kruke (2017), "it is essential that preparedness follows an all-hazards logic based on uncertainty and the unexpected, rather than the expected and routine". Experts and practitioners agree that the first few hours are crucial. The longer the wait, the more energy it takes to regain control and the more impactful the decisions will be.

Territorial Crisis Management,
coordinated by Richard LAGANIER and Yyvette VEYRET. © ISTE Ltd 2022.

The first challenge facing crisis management is that of continuity and consistency with risk prevention. International actions and recommendations have evolved from an almost exclusive focus on crisis management and post-crisis assistance ("relief") to a broader focus on coherence with risk prevention in the framework of DRR (disaster risk reduction). Even if progress has been made, the increasing cost of major disasters over the last 20 years and their deep and lasting territorial impacts (Katrina 2005, floods in Central Europe 2002, tsunami in South-East Asia in 2004, earthquake in Haiti in 2010, tsunami and nuclear accident in Fukushima in 2011, etc.) have shown the limits of an approach to disaster reduction that is based solely on crisis management. There is no point in constantly spending energy and money to assist disaster-stricken countries if there is no integrated and comprehensive disaster risk reduction strategy at its core.

The second challenge of crisis management is certainly political. How can we involve all the players in crisis management? How can all the stakeholders be brought together? The evolution tends towards a greater involvement of the exposed populations who are the "first actors of their security". The disaster victims are the first ones on the spot and for good reason! During disasters, we see inhabitants coming to the aid of disaster victims, even before the arrival of rescue professionals. In spite of this, the involvement of inhabitants or citizens (future disaster victims!) in the preparation of the crisis is not obvious for different reasons. The need for efficient crisis management therefore requires, in addition to the involvement of institutional and socioeconomic actors, the development of new forms of involvement for the people exposed.

The third challenge is the association of the technical and humans. The field of crisis management has broadened from the practices of rescue professionals, which have existed for centuries, to the development of expertise and engineering upstream and downstream of crisis management itself. This engineering of crisis management in the broadest sense of the term includes monitoring, forecasting, anticipation, including planning, as well as post-crisis management and even reconstruction and the return to normalcy with all the nuances that this formula implies. Managing crisis means ensuring the link between growing technical expertise and the need to involve as many actors as possible, first and foremost the disaster victims themselves. As Alexander (1997) already pointed out, the increase in technicality and the disciplinary or specialized approach to crisis management has been encouraged by the professional evaluation methods of the services in charge of crisis, internal evaluation methods by peers who judge the degree of technicality, and the mastery of procedures and techniques deployed during crisis management. It is done at the expense of a global and integrated vision of crisis (Kachali et al. 2018). Thus, above applying tools and procedures, it is a matter of focusing our attention not on specific crisis management procedures and tools and their compliance, but on the control of the context in the sense that Alexander (1997) defined it. Thus, the questions that will guide us are to identify the conditions and levers that allow us to prevent a situation from degenerating into a crisis, avoid going from a crisis to a catastrophe or try to quickly regain control of the situation.

2.1. Nature of the crisis and forms of crisis management

2.1.1. *The nature of the crisis: cycle, trajectory or "black swan"?*

The previous chapter examined the diversity of crises that our societies may face. The characteristics of these crises influence the way in which they are managed. The duration of the crisis, its spatial extension, its predictability (flooding vs. chemical explosion), its holistic or sectoral character, etc., are all parameters that influence the management of the crisis and therefore its anticipation. Another question prior to the study of crisis management is the search for meaning in the crisis. How does such and such a crisis make sense? Meaning as significance, that is, what does it reveal? But also meaning as a progression, a direction as in the expression "sense of history". What is its place in the process of evolution of societies and their environment? Are the crises part of a cycle that repeats endlessly or are they stages in an evolution, in a trajectory?

The cyclical vision of the risk and crisis binomial has been widely criticized (Pigeon 2012; Moatty 2015). It presents a restrictive vision of crisis as an inevitable stage in the vicious circle of prevention, risk accumulation, disasters, reconstruction, risk oblivion and new crisis. It implies that it is not possible to avoid crises and that they are inscribed in the territories, repeating themselves in the same way, in the same places, etc. However, the causes of crises evolve. Some disappear, others appear. Indeed, for a good number of crises, the deep causes are external, independent of the source of danger and have their deep roots in the evolution of societies and territories. The 2003 heat wave, which caused 70,000 deaths in Western Europe, is the symbol of the convergence of several independent developments: atmospheric warming, the aging of populations and the problem of how to deal with it, and changes in the health system.

Thus, is the crisis the expression of the unusually high intensity of a known phenomenon: a 100-year flood compared to a 10-year flood, for example? A category five cyclone compared to a category two or three cyclone? Or is it of a new kind that societies have not experienced in this form? Taleb (2010) refers to these so-called “black swans” as “unpredictable” events, beyond the reach of return period calculations and whose consequences are far greater than what can be predicted by existing models. It especially evokes financial phenomena (of the stock market crash type), but the reasoning is applicable to the so-called “natural” or technological risks. The unforeseen is inescapable if we consider the evolution of socioeconomic and natural systems. Crisis management plans must allow for a certain resilience understood as “the capacity to cope with unanticipated dangers after they have manifested” (Wildavsky 1991). Thus, it is necessary to maintain a certain distance from the usual phases of a crisis: watch, vigilance, alert, safeguard, relief, post-crisis, rehabilitation, reconstruction, etc.

2.1.2. *The nature of the crisis: scale of the crisis versus scale of its management*

2.1.2.1. *Contraction of time and dilation of space*

Crisis managers are used to saying that a crisis is characterized by a contraction of time and a dilation of space. Time is accelerating and decisions have to be made in ever shorter time frames. Moreover, due to – among other things – accessibility difficulties, space expands: travel speeds are slower, communication routes are unusable (railroads, roads, airports, etc.) and information feedback is more difficult.

2.1.2.2. *Disaster areas and crisis areas*

The territorial dimension of crises varies greatly, from intense but very localized crises (earthquakes) to more widespread or even global crises such as pandemics. Crisis management also has its own territorial logics which are not those of the crisis itself. Crisis management becomes problematic when the damage affects the very resources of crisis management. The earthquake in Haiti in January 2010 was catastrophic in that it affected the entirety of a small state, including its capital, Port-au-Prince. Beyond the lack of resilience linked to chronic poverty, the small size of the Caribbean micro-territories reinforces their vulnerability to natural phenomena such as hurricanes, tsunamis and earthquakes. For Haiti, if the aid came from outside, the difficulties of coordination would have suffered even more from the absence of a local power, completely disorganized by the crisis. Moreover, the influx of external aid increases the political dependence of these island microstates, in this case with respect to the neighboring United States. The case of Haiti will be compared to that of Saint-Martin and Saint-Barthélemy, which benefited from resources projected from the French mainland following Hurricane Irma. However, for certain risks, such as epidemics, insularity can be a factor favoring the territorial control of the crisis (regulation of arrivals in the territory). The only areas that escaped the Spanish flu in 1918–1919 were remote and isolated islands that were able to implement effective quarantine measures in time.

2.1.2.3. *Crisis territories and crisis management territories*

The debate between supra versus local crisis management is a false question. During the Covid-19 crisis, for example, the merits of states with decentralized structures were praised (German federalism, which gives a great deal of latitude to the Landers). This management close to the local territories would guarantee greater responsiveness and adaptation to the local context through targeted measures. But the pandemic is by definition global, and we saw the limits of a localist approach in Switzerland during the second phase of the epidemic in November 2020, when each of the 26 cantons enacted its own prevention measures. As the business closure measures did not affect the same activities, the citizens of one canton crossed the border of their canton to benefit from a business or a service that was not closed in the neighboring canton. This quickly led to an increase in travel and the spread of the virus.

The interweaving of jurisdictions, which goes hand in hand with the multiplicity of institutional actors in crisis management, poses the problem of the decision-making scale. The question of crisis management actors and the clear definition of competences is a common theme in the evaluation of crisis management systems, but it is not an abstraction. Conflicts of actors often cover conflicts of competences. Who does what and where? Crisis management has territorial stakes. It can be used to assert power over a territory. Let us take the example of crisis management in Saint-Martin during the passage of Hurricane Irma in 2017. The prefecture of Saint-Martin in the French West Indies is a delegated prefecture dependent on the prefecture of Guadeloupe. This dependence poses a problem in the pre-positioning of resources during cyclonic crises because Guadeloupe and the northern islands are likely to be affected by the same cyclone. The accuracy of the trajectories of cyclones, which *ultimately* determines whether one or the other will be affected, is too late for resources (which would deprive Guadeloupe) to be projected on Saint-Martin. In addition, the regional level (EMIZA: *état-major interministériel de zone Antilles*, interministerial zone staff of the West Indies) includes Martinique, Guadeloupe and the northern islands, and is headquartered in Martinique. It was marginalized in 2017 when the prefecture of Guadeloupe and the Cogic in Paris took leadership of crisis management.

2.1.2.4. *The ordinary, the crisis and the domino effect: the management of long crises*

Crisis management plans are often based on a hazard-centric and single-area approach. They hardly consider the possibilities of domino effects. By domino effect, we mean "an amplification of the initial crisis linked either to the occurrence of a new hazard (a tsunami after an earthquake) or to systemic effects with cascading failures within the crisis management organization (an epidemic and a failure of the healthcare system)" (Daudé et al. 2021). These domino effects are difficult to predict. Their onset makes plans more complex and sometimes leads to contradictory instructions. However, during long crises, this possibility can be frequent. During the Covid-19 pandemic that started in 2020, many local crises appeared. It was therefore necessary to manage the crises within the crisis. The complexity of managing Natsan (Natural and Health) or Natech (Natural and Technological) crises has been highlighted by many authors (Piatyszech et al. 2017a).

The management of long crises requires dealing with several constraints:

– the territorial variability of the disaster areas. The locations of the crisis move within the same risk territory;

– the reconciliation of priorities: economic, ethical, political, social. When the crisis becomes long lasting, priorities (economic, political, etc.) that had been put on hold for a while and the day-to-day functioning of businesses resurface. The actors of the management of the territories must put their activity continuity plans into action, in particular the managers of networks and essential services (water, electricity, food, etc.);

– the interweaving of crises with the appearance of local domino effects (Daudé et al. 2020);

– the evolution of knowledge and the adaptation of decisions.

At the other end of the frequency scale, ordinary losses tend to be masked by major crises. The crisis tends to be seen as a "total" event that focuses all attention and resources. Examples include the 1993 Mississippi River Basin floods that migrated downstream in five months, the fires in Australia between September 2019 and January 2020, or the Covid-19 pandemic in 2020–2021. Thus, emergency and health services are focused on the management of the major crisis, thus neglecting the daily disaster situation. The management of ordinary claims (road accidents, domestic accidents, chronic illnesses, etc.) may be impaired as a result.

2.2. Preparing for the crisis: the prevention/preparation link

The concepts related to crisis management preparedness (planning, preparation, monitoring, etc.) are still not very stable. Staupe-Delgado and Kruke (2017) give a definition of *preparedness* on two levels: a minimal level that could be summarized by an active, continuous and anticipatory preparation of the crisis and a broader level including a social dimension. Indeed, social capital, perceptions and cultural traits influence the state of preparedness of a territory. Moreover, this component includes the mobilization of populations, that is, the involvement of populations, whereas a restrictive vision of crisis management preparedness will see populations as a problem to be managed, or a source of disorder to be controlled. Thus, if "readiness" is a continuous training process that relies primarily on planning, it covers a much broader field that begins with a strong link and consistency with prevention.

2.2.1. *Preparation: the imperative link between prevention and crisis management*

Crisis management and risk prevention have long been disconnected. This disconnection is multi-causal. It can be explained first of all by a separation of the actors and authorities in charge of the two fields. Crisis management is the responsibility of the emergency services, possibly the army and/or the police. Risk prevention is the responsibility of government agencies (health, environment) or local authorities. In France, for example, natural hazard prevention is the responsibility of the Ministry of Ecological Transition, while crisis management is the responsibility of the Ministry of the Interior.

At the same time, it is conceivable that it would be difficult to establish a "Ministry of Risk and Crises" unless an omnipotent ministry was to be decreed, which would quickly become ineffective. The regrouping of risk and disaster prevention under the aegis of a single agency is illusory. It is a question of sharing culture, sharing information and coordination. On the other hand, placing crisis management at the national level directly under the authority of the Prime Minister or the President of the Council, as in Italy, avoids a tug-of-war between ministries.

Risk prevention is based on an arsenal of measures spread over the long term: preventive vaccination has reduced the risk of an epidemic crisis even if it has not eliminated it. Taking into account a certain number of measures in the planning of the territories allows us to minimize the consequences of a crisis or facilitate its management. For example, the design of buildings can facilitate the evacuation of people or accelerate their restoration in the event of flooding. The probability of risk also changes under the effect of exogenous factors that are not linked to voluntary and specific prevention actions: the warming of the atmosphere reduces the risk linked to extreme cold. The improvement of housing in rich countries increases the resilience of territories, while the precariousness of housing is a growing factor of vulnerability in poor countries.

Preparedness (for a crisis) involves actions to anticipate a crisis and deal with the residual risk not covered by prevention (Staupe-Delgado and Kruke 2017). It therefore involves accepting that the risk will not be eliminated and living with it. The link between prevention and preparedness raises the problem of reference events that will serve as a basis for prevention, on the one hand, and crisis management planning, on the other. If we admit that crisis management deals with the residual risk that could not be reduced by prevention, then the standards are different and complementary. But how can the acceptable level of risk be set? It depends on the type of hazard and in particular on the degree of control of the danger, as well as on the acceptability (financial, political, social) of the risk, and the territorial context. Preventive measures, politically or financially too costly to implement beyond a certain return period or with too much uncertainty, give way to crisis management planning. The hazard scenarios taken into account for technological risks are 100–1,000 times rarer than those for natural risks because technological risk is supposed to be controllable by humans. The requirements are lower for natural risks because an uncontrolled part of the hazard is admitted (cyclone, for example). If we reason in terms of return period, the 100-year event is retained in France as the reference event for prevention, particularly for floods. We may be tempted to say that crisis management takes charge of the over 100-year event not covered by prevention. In reality, prevention is not 100% effective, and there is still a residual risk in the 100-year envelope. Better (or worse!), the 100-year flood envelope is often used as a reference for prevention and crisis management planning. Intermediate risk levels with a more frequent return period (Q20 or Q30) are not always taken into account, nor are high intensity and very low frequency events (Q200 and beyond). The European directive of 2007 obliged the managers of highly exposed areas to provide flood maps for return periods below and above the 100-year flood, implicitly considered as "average" in the directive. The link between emergency plans and prevention is beginning to emerge, but it is far from being systematic.

Prevention must reduce the risk of a crisis to a certain level. Thus, crisis management must deal with what is not contained by prevention measures. The definition of prevention measures could take into account the damage avoided by efficient crisis management, in order to avoid prescribing very severe measures as in rare and very damaging events. The coherence between prevention and crisis management would be promoted by a more systematic evaluation of the costs and damages avoided. Structural prevention measures have long been the subject of more or less reliable cost/benefit assessments. But non-structural measures (including crisis management) are rarely evaluated. This lack of evaluation is due to technical constraints (lack of data, lack of methods) as well as ethical constraints (how much is a life saved worth? How can the loss of value on the environment be evaluated?) without forgetting the question of the temporal horizons to be taken into account. This economic evaluation of both costs and avoided damages is nevertheless requested by the emergency services anxious to justify their function and by their financiers (state, local authorities). Evaluations exist on a case study basis, but they are far from being systematized. Penning-Rowsell and Wilson (2006) estimated the cost of crisis management during the autumn 2000 floods in England at 15% of the total damage. The evaluation of the cost of crisis management, as well as that of avoided damage, would provide elements for a decision to relate these costs to those of prevention.

2.2.2. *Monitoring: choosing the right indicators*

On the borderline between prevention and crisis preparedness, expert monitoring consists of detecting weak signals that may indicate a future increase in perils and possible crises. This watch concerns sources of danger (natural, technological, terrorism, etc.) and emerging risks such as future pathogens. But this watch may also address the organizational and social systems likely to evolve and generate future crises. The aging of populations in European countries, as well as in so-called emerging countries, can be a factor of future crisis due to the increase in human and social vulnerabilities that it induces. Similarly, global warming, rather than being a risk in itself, is a factor in the amplification of certain hazards such as heat waves or rising sea levels. Detecting socioeconomic, demographic and environmental changes that may cause a crisis later on is part of crisis management preparation. Detecting weak signals is, however, complex: it is necessary to be able to spot them, drowned in the noise of information, know how to interpret them and be able to transmit them to the decision-making level, which supposes that the sorting out escapes psychological and organizational biases.

2.2.3. *From monitoring to forecasting*

The increase in monitoring leads to vigilance, which can be translated into forecasting or prediction when the technical tools allow it. Vigilance systems at all levels, from local to international, provide information on the evolution of a risk likely to degenerate into a disaster and/or crisis[1]. The differences between foreseeable and unforeseeable risks obviously introduce an essential distortion in crisis anticipation (Leone et al. 2010). Earthquakes can be considered foreseeable, but not predictable. Some areas are known to be highly susceptible to hazards, but predicting earthquakes in the near future is not possible, even if certain warning signs sometimes allow it.

For earthquakes, we know the areas where they are likely to occur, but we cannot say when they will occur. The interest of prediction for earthquakes is questionable: a seismic construction strategy, which avoids most of the destruction and thus preserves lives (an earthquake does not kill; it is the collapsing building that kills), is much more effective and much less costly in human and economic terms than evacuation. Apart from earthquakes, most natural phenomena are predictable, in the near future.

2.2.4. *Forecasting tools*

Forecasting allows us to anticipate events and thus gives us time to prepare crisis management. Depending on the type of event that triggers the crisis, forecasting may or may not be possible.

To be able to predict, we must have data: historical data first of all. Most current disasters are reproductions of past disasters: for example, the Dracénie flood of June 2010 is an almost identical reproduction of the one of July 1827. The study of archives allows us to identify vulnerabilities and give references for the implementation of observation networks and modeling. Data from an observation and monitoring network are added to historical data. The density of the network and the continuity of observations over time are essential for the understanding of phenomena and the reliability of forecasts: meteorological data (wind, storms, forest fires), hydro-meteorological data (floods, avalanches), observation of land movements (landslides, volcanism, etc.). Space or airborne observation data will be included.

Forecasting and prediction are based on models, among other things. A model is a numerical tool that attempts to reproduce reality. Based on the known laws of formation of phenomena (laws of physics, for example, such as fluid mechanics, thermodynamics, etc.), its conception is more or less rustic, according to the available observations and the use that we want to make of it. The model is parameterized and adjusted by integrating previous events. There are countless

1. See: https://www.gdacs.org/default.aspx.

examples of predictive modeling, from simple rainfall/flow models to epidemic propagation models. When the forecasting/prediction tools are not very robust or when the risk that appears is poorly known (new virus, for example), we proceed by analogy using knowledge or tools calibrated on similar risks. This was the case with Covid-19, for which epidemiologists used models of influenza or SARS 2003, which were better informed, but which obviously led to false leads and uncertainties, as incubation times and contagiousness were different. During the chikungunya crisis in Reunion Island in 2005–2006, the health authorities used the dengue guidelines in the first months, which led to damaging decisions with prevention based solely on vector control to the detriment of social or behavioral considerations (Metzger 2009). The disease was well known but in "developing countries". It should be emphasized that imperfect references are better than no reference at all, as was the case during the 2003 heat wave in Europe.

All forecasts are subject to uncertainty. In general, the closer the forecast is to the event, the smaller the margin of uncertainty is, because reliable data are injected into the model. For example, for a cyclone, when it is still far away, the margin of error on its trajectory is about 100 kilometers per 12 hours, or 200 kilometers when it is 500 kilometers away, with the cyclone moving at an average speed of 20 km/h. The margin of error is reduced to 30 to 60 kilometers less than 24 hours in advance, which is ultimately small compared to the size of these phenomena, from 200 to 800 kilometers in diameter. Forecasting centers such as the National Hurricane Center in Miami publish cones of uncertainty around the probable trajectories of cyclones. The reduction of these uncertainties generally requires the comparison of different mathematical models and human expertise. The taking into account of these margins of error by decision-makers in crisis management is the result of both expert opinion (cost–benefit analysis of taking into account the different scenarios in particular) and political decision; there is a major issue of communication on these uncertainties, both for decision-makers and the general public. The translation of data from forecasts in the crisis room into operational terms is an imperative, which may require the use of "popularizers". Thus, in France, RDIs (*référents départementaux inondations*, departmental flood point of contacts) have been introduced with departmental prefects.

2.3. Information in times of crisis

Information is an essential issue in crisis management. This includes upstream information, that is, real-time monitoring of the situation, as well as communication in the form of so-called "downstream" alerts, that is, addressed to those exposed. In the event of a crisis, the risks of confusion in information, interference between actors (particularly between experts and managers), excessive information and false news are forms of loss of control that sometimes turn out to be the source of crises or their amplification.

2.3.1. *Upstream information in times of crisis: monitoring tools*

"Real-time" information tools have developed considerably. They sometimes differ from vigilance tools because the monitoring of the phenomenon at the origin of the crisis (meteorological, etc.) is added to or replaced by the monitoring of its effects (damage) and of the crisis management itself (allocation of resources, population movements, etc.). Crowdsourcing and crisis-mapping are interesting contributions in times of crisis. They refer to the collection, production and free availability of useful information for crisis management (Douvinet et al. 2018). In crisis-mapping, whose best known tool, OpenStreetMap (developed in 2004), is not only used during crises, the mobilization of a community of cartographers provides online information and decision-making tools to rescue teams, authorities or NGOs involved in crisis management. This was the case after the Haiti earthquake in January 2010, where the mapping of the disaster area was reconstructed in a few days from aerial photographs (Ghosh et al. 2011).

Information feedback from the field is a key point in crisis management. Even if tools have been developed, this research is sometimes difficult. Crisis managers must identify ex ante the most relevant indicators for monitoring the crisis. These data must have four qualities: diversity, robustness and reliability, speed, and proximity.

Information feedback should not depend on a single source. For floods, official messages from national monitoring networks, when they exist, must be doubled and are often duplicated in the territories by informal networks between downstream and upstream municipalities, for example (Creton-Cazenave 2010).

The multiplication and combination of the means and sources of information for monitoring favors the robustness of the system of monitoring and information. The robustness of monitoring data is due to the experience of the sender, as well as the transparency of the data and methods. An example of this is the suspicion that arose during the 2020–2021 pandemic about the number of Covid-19[2] victims in certain countries, for political or technical reasons.

2. See: https://coronavirus.jhu.edu/map.html (Johns Hopkins University Epidemiology Reference Site).

The speed of implementation is also a key consideration. During the Covid-19 outbreak in 2020–2021, the detection of viral material in wastewater by the Obepine network proved to be an interesting indicator, relatively easy to implement, inexpensive, quasi-instantaneous and "synthetic". It gave a vision of the progression of the infection at the scale of the housing block that PCR tests do not allow at this spatial scale. Similarly, the monitoring of territorial crises uses remote sensing tools (satellite or airborne) (Klemas 2015). They are at the heart of the European Copernicus[3] provision, which can collect and analyze satellite data rapidly and make it available to crisis managers free of charge.

The final requirement of the information tracking system is to maintain a link to the field. A counter-example illustrates this need. During the 2003 heat wave, although the first mass deaths occurred as early as August 4, the White Plan was only activated by the government on August 13. It is true that most of the political leaders were on vacation. But this lack of reaction can also be explained by the absence of rapid awareness of the problem. The epidemiological data that came back from the state services were partial and arrived on the minister's desk with delay. The filtered, averaged and abstract statistics did not allow for a concrete vision of the reality. However, as early as August 7, the emergency services were reporting a massive influx of elderly people with hyperthermia. Likewise, the *Pompes Funèbres Générales*[4], which had an almost daily count of the number of deaths, was able to estimate the excess mortality at 500 per week as early as August 12. A quick visit by the minister or an official to a morgue or hospital emergency department would undoubtedly have precipitated their awareness and reaction. Even if this would probably not have prevented a significant number of deaths, the health crisis might not have generated such a political and media crisis (resignation of the Minister of Health…).

The feedback and results of experiments are, to say the least, contrasted between the advocates of "more information generates better decisions" and those who, on the contrary, emphasize that, in order to make rapid decisions, the complexity of information must be reduced (November et al. 2020).

2.3.2. *Alert*

An alert, that is, the warning of an imminent danger to exposed populations, moves the event from the expert domain to the political domain, because an alert is above all a political act that is the responsibility of the various authorities in charge

3. See: https://emergency.copernicus.eu/mapping/list-of-activations-rapid.

4. Funeral directors.

of crisis management. This is a key point in crisis management that has given rise to a great deal of research (Douvinet 2020).

2.3.2.1. *From warning means to warning system*

Two major developments have marked the evolution of warning over the last 20 years. The multiplication of the means of detection and dissemination of information has doubled official alerts. Warning channels have multiplied with the development of the Internet, social media, scientific and technical communities (amateur meteorologists, for example), specialized private service providers using public data for forecasting, etc. There is thus confusion between alerts and warnings. Of course, this phenomenon is not new. In the past, local residents could be surprised by a flood even before being "officially" alerted, and informal alerts were often more rapid than official alerts for fast-moving phenomena (Creton-Cazenave 2010).

Thus, the authorities in charge of the "official" warnings must think about a more efficient warning system combining several warning means by integrating recent technologies (Douvinet 2020). The reasoning of national or local authorities in terms of warning has often been a specific reasoning on one or several means of warning[5] (e.g. siren or telephone), without thinking about an integrated and coordinated warning system where the different means are implemented to reach the whole population and actors. French authorities persist in keeping the costly and inefficient siren system, while countries like Belgium have abandoned it (Douvinet et al. 2019). We are also sometimes surprised by the delay in the use of *cell broadcast* for which ethical, economic and even political obstacles are numerous. But the European Union is about to impose the use of cell broadcast as an alarm system.

2.3.2.2. *Message content*

Beyond the warning system, the form and content of the warning message are major issues. They form a difficult equation between the need to be vigilant and to mobilize without "frightening", between stating the reality and attenuating it to avoid panic effects. The will of political actors and managers to show that everything is under control can lead to damaging exaggerations, semantic contortions and even lies in certain situations, as was verified a posteriori concerning the limits of the radioactive cloud that spread towards Western Europe following the Chernobyl nuclear accident.

5. Following the Swiss cheese model, it is a matter of adding warning resources like stacking layers of cheese in the hope that the holes in the bottom layer will be plugged by the top layer (Stauffer et al. 2017).

The normative approach of issuing general instructions "blindly" in crisis management has two advantages: it is convenient because it is based on existing rules, a legal arsenal and known procedures. The other advantage is that of equality regarding the law. However, instructions are adapted to the norm and may leave out those who deviate from the norm. Can we ask a disabled person in a wheelchair, a person weighing 120 kg or a tall person to take refuge under a table in case of an earthquake? In the same way, instructions sometimes impose conscious or unexpected choices between contradictory instructions. During the Xynthia flooding in 2010 in the west of France, the advice was to confine ourselves: "stay at home". This instruction, relevant for winds, precipitated the loss of people who lived in coastal areas subject to flooding and who should have sought safety in upper stories. Also noteworthy are the cases of instructions that are valid to a degree. In 2015, eight people died in the flooding of an underground parking lot in Mandelieu-La-Napoule in the south of France. Residents used to go looking for their vehicles to park them in a safe place during flood warnings. For years, the inhabitants thus followed an instruction to protect their property, which turned into a trap when a flood rose faster and higher than usual. The normative approach has its limits, and the need for idiosyncratic adaptation and territorialized appropriation of the instruction is measured (Gaillard 2019).

2.3.2.3. *Politics and crisis*

The scope of warning and instructions and trust in the issuing actors are intimately linked (Mendenhall 2020). The instruction will be more respected if it is "fair", that is, clear, adapted to the situation and the public, timely in terms of chronology and proportionate. In return, the credibility of alert issuers is linked to appropriate communication. Poorly controlled crisis communication can ruin well-prepared crisis management, and the inability of the authorities to deal with the crisis has often resulted in political failure. The preparation of political leaders and authorities in charge of crisis management in communication, leadership during the management of major crises out of the framework (e.g. Katrina, Fukushima), its role and its influence on the decisions taken (Boin and Hart 2006; Laganier 2015). Zamoun et al. (2018) emphasize the importance of contact between crisis management leaders, agents and the population, especially during long crises that must inevitably be paced.

The communication of political leaders during the Covid-19 crisis has been widely commented upon. For example, the announcement of the containment by Jacinda Arden and Boris Johnson was compared. The New Zealand Prime Minister gave a press conference on March 23, 2020, focusing on three themes: giving direction, explaining and showing empathy[6]. The conference was followed by a lengthy question and answer session from the press. At the same time, the British Prime Minister's message was prerecorded in a very "official" setting with no interaction with the press and a commanding speech using the words "instructions" "the police have the power to enforce the rules...". Arden applied Mayfield's rules (2015, 2017), according to which the communication of leaders in a crisis must give direction (direction-giving), make sense (meaning-making) and show empathy.

2.3.3. *Infodemia, misinformation and controversy*

Sometimes, information is in excess with difficult arbitrations and uncontrollable drifts (fake news, manipulations, contradictory expertise and scientific controversies, etc.). The word "infodemia" was born about 20 years ago during the SARS epidemic in 2003, which saw the uncontrolled appearance of false information with the development of the Internet, which has multiplied the dissemination of information of any kind. The circulation of false news, scientific and/or political controversies is even part of the crisis itself[7]. It is not new and did not appear with the Internet or with social media. It is a parameter to be integrated for managers. We would be tempted to say that controversy and even conflict are consubstantial to the crisis. A crisis "that goes well" is not a crisis. In the face of the unknown, scientific controversies emerge on the origin of the risk and the means to combat it. The controversies over the effectiveness of hydroxychloroquine during the Covid-19 crisis in 2020–2021 are reminiscent of the scientific controversies that erupted during the Spanish flu in 1918–1919 over the origin of the disease (bacillus or virus) (Rasmussen 2007). This raises the question of the status of scientific knowledge and its use in crisis management. The time of crisis management is not the time of scientific expertise. The latter requires time for verification. However, crisis managers require information as quickly as possible. For long-term crises, especially pandemics, the status of risk forecasting expertise is changing. We are moving from the status of the uncontested expert to a kind of knowledge that is discussed, put in the public arena, more transparent in a sense, but where uncertainty is displayed. This evolution was clearly manifested following the Aquila earthquake in 2009, which saw the condemnation of scientific experts by the justice system, and during the Covid-19 pandemic, during which numerous scientific controversies were made public.

---

6. See: https://www.tellerreport.com/news/2020-03-14---comment--when-the-world-gets-infected-with-the-virus--a-dangerous-infodem-also-spreads-.BklCgzScS8.html.

7. See: https://informationdemocracy.org/2020/11/12/250-recommendations-on-how-to-stop-infodemics/.

Crisis management can get out of hand when politics (in the political sense) interferes. There are many examples of political recuperation, such as the closing of borders during the Covid-19 crisis. We also note the attempts to discredit certain whistleblowers suspected of political intentions, as was the case for the first doctors who denounced the extent of the Covid-19 epidemic in China in January 2020.

2.4. Crisis management: an integrated approach

Crisis management has become both professionalized and expanded. We have gone from an empirical crisis management based on experience – in particular of emergency services and public services in general – in a *top-down* relationship with the welfare state, to the need to organize a planned crisis management, integrating more and more actors and more and more complex crises. Because of their territorial complexity, crisis situations involve a large number of actors with varied skills and resources, and above all, function according to various organizational logics (hierarchical, deliberative, individual or collective decision-making, etc.) and obey various rationales (public, semi-public and private actors). There are therefore important issues concerning the distribution of competences between actors (who does what? In relation to whom? On what occasions?), the distribution of capacities and decision-making modes, the management and mobilization of available resources, the circulation of information, etc.

2.4.1. *Conditions for a shared vision*

2.4.1.1. *The representation of impacts*

The representation of the potential impacts of a disaster is an essential prerequisite for its proper management. It is a matter of putting the disaster in the realm of possibility and making a shared diagnosis of crisis scenarios. During the Xynthia flooding episode in 2010 in the west of France, neither the political leaders nor the population, nor the emergency services had the damage that could be caused by a seasurge in mind. Of course, it was known in the vernacular culture or among some state engineers that certain areas were submersible, but between a vague mapping of the hazard and the detailed knowledge of the potentially submersible areas, the number and type of houses, the type of population, the road access, the possibilities of evacuation and refuge, etc. There is a gap that the English and Dutch had partially filled, because they were sensitized by the catastrophic event of February 1, 1953, which caused more than 300 deaths in Great Britain and 1,850 in the Netherlands (Lumbroso et al. 2011). The necessary mental representation of impacts also extends to the consequences of prevention or safeguarding measures that may be deployed. In complex and multiphase crises such as the Covid-19 epidemic in 2020–2021, tradeoffs are frequent between the benefit of prophylactic measures and their socioeconomic effects, sometimes with unexpected impacts (closing schools limits the spread of the epidemic, but handicaps economic life and harms children in terms of schooling and sometimes psychological health).

Numerous tools participate in the development of these impact representations, from simple cartography to multi-agent models and video. Examples include the Life Safety Model© and the Flood Risk to People© developed several years ago by British teams to model the impacts of flood protection structure failures (Brussee et al. 2021).

"Anticipation means problematizing the future" (November et al. 2020). It is a matter of sharing a vision of "what lies ahead" between crisis management actors (including exposed populations). This representation gives meaning not only to crisis preparation, but also to prevention. The question then arises as to the reference event, the scenario or scenarios chosen: the past, known and plausible event, the frequent and relatively low intensity event which has the advantage of a not too distant time horizon, or the major, catastrophic event which will cause significant damage, but whose representation may suffer from a lack of credibility and whose distant time horizon may demobilize. Another underrepresented example is the impact of cybercrime and, more generally, of the breakdown of computer networks. After the fire that destroyed the OVH company computer servers in Strasbourg on March 10, 2021, many websites and companies were left without any support. On Meteociel, a collaborative portal that provides many services and weather information, we could read: "this is the most difficult moment since the beginning of Meteociel 18 years ago, we have had a few breakdowns but we never expected to see a *data center* disappear like this".

2.4.1.2. *Coordination and common culture between actors*

This representation of impacts serves as a basis for a shared culture of crisis by managers and as a cognitive reference for the populations at risk. It is the necessary condition for good coordination between actors. Feedback from experience still stigmatizes poor coordination or even interinstitutional or interpersonal conflicts in crisis management. The poor prior definition of the roles and status of each person is at the root of the dysfunctions. We have already mentioned those linked to the interlocking scales of crisis management actors, with attempts to appropriate management by one level or another to the detriment of inter-service cooperation.

Beyond this lack of institutional coordination, effective crisis management requires "a shared vision within and between crisis units on the nature, time horizon and methodological framework of the anticipation function" (November et al. 2020). The different actors at different territorial levels have different operating modes and objectives. During a crisis, a mayor's request, for example, will be to be able to benefit from the rescue forces as soon as possible in their commune, whereas the objective of the rescue services will be to distribute their forces to the places where they are needed. The question of the multiplicity of actors is no longer a problem if they share common crisis management objectives. In a survey of crisis management departments in Europe (Lumbroso et al. 2011), the question asked to crisis managers was: what do you consider to be the crucial factor for successful crisis management? Before the quality of monitoring, mapping, knowledge of the hazard or the abundance of resources, it appeared that the crucial factor for successful crisis management was the inter-knowledge and inter-understanding of the actors. The following chapter (Chapter 3) looks at the learning methods that can promote this inter-knowledge. Exercises are one example. In the West Indies, the "Richter" exercise[8] (preparation for a major earthquake) and the Caribe Wave[9] exercise (preparation for a major tsunami) have been beneficial even for the preparation of hurricane crises. A Richter exercise took place in March 2017. It allowed the crisis managers of the Caribbean area to know their interlocutors, modes of operation and needs, and therefore save time during the hurricane crisis Irma a few months later.

2.4.2. *Integration of populations*

The role of states in the management of major crises is important. We saw this with the Covid-19 crisis which, far from highlighting international cooperation, showed the primacy of the national framework in crisis management (Boniface 2020). International competition for crisis management resources (masks, vaccines) is sometimes strong, even if states that are too weak are not able to face major crises, as we saw in Haiti in January 2010. Rather than wanting to manage everything at all costs, the role of the state is above all to allow everyone to exercise their responsibilities. The state does not have the vocation, nor the possibility of substituting itself for all actors in crisis management. Also, the involvement of

8. See: https://www.guadeloupe.gouv.fr/Politiques-publiques/Prevention-securite-des-personnes-et-des-biens/Defense-et-protection-civiles/Exercices/EU-Richter-Caraibes-2017-un-exercice-de-crise-de-grande-ampleur-aux-Antilles.

9. See: https://www.tsunamizone.org/caribewave/.

populations in crisis management is certainly a problem of coordination, trust and communication to insert local initiatives in crisis management.

2.4.2.1. *Involvement of populations and actors*

The involvement of the population in the preparation and management of a crisis is also a crucial issue that can be based on the need to develop risk awareness prior to any acquisition of crisis culture, crisis culture that can be manifested by the knowledge of appropriate behaviors. This involvement of the citizen "actor of their security" and local communities is a requirement in international recommendations (UNISDR 2015). The resilience of individuals and communities must complement the commitment of institutional actors in crisis management. Residents are the first responders during a disaster, and Kruke (2015) points out that "preparedness has not sufficiently involved local people arriving on the scene of first an event, neither in the planning process, nor in the response and recovery phases".

This interest in involving the inhabitants is not self-evident for public authorities, who feel that their protective role requires them to reassure at all costs and show that everything is under control, whereas, in order to build trust, it is necessary to tell the truth, including the gaps and weaknesses. This uncomfortable posture of truthfulness also allows everyone to exercise their own judgment in order to avoid, at the very least, putting themselves in danger and aggravating the crisis. However, this assessment of the authorities' cautious attitude should be qualified. Since the wave of terrorist attacks in France in 2015, many measures involving civil society have been reactivated, such as the special security plans for schools. The government website "R-SQUES, *prévention des risques majeurs*[10]" encourages volunteer work and commitment in the event of a crisis: becoming a reservist (health reserve, military reserve, gendarmerie reserve, civic reserve, cyber-defense citizen reserve, etc.), being a volunteer firefighter (almost 200,000 people in France), or more simply: donating blood, training in first aid or adopting good digital practices. On this last point, there are a number of tips: chase rumors, do not use your cell phone if not necessary, when your safety is ensured, report the elements of your environment that may be relevant for rescue on social media. The association VISOV – *Volontaires internationaux en soutien opérationnel virtuel*[11] Operational Support – intervenes in the technical or methodological support of a crisis management organization. It aggregates the information available on social media and ensures its massive sharing in order to support the disaster-stricken populations and emergency services with which partnerships are signed.

10. See: https://www.jeunes.gouv.fr/R-SQUES-prevention-des-risques.

11. See: http://www.visov.org/.

As far as the inhabitants are concerned, the knowledge of the possible crisis scenarios, and the understanding of the validity of the measures requested by the authorities facilitates appropriate behaviors, even timely initiatives. However, a real difficulty for residents and non-specialists is the very nature of disaster risk. Before its occurrence, it remains a theoretical notion, an event that we do not know in advance and that will have dramatic consequences. The knowledge of past events, of possible scenarios, is necessary, but it is not sufficient. The numerous psycho-social experiments (since 1957 by their initiator Festinger), which have led to the development of theories on the influence of knowledge or ideology on behavior, have shown that it is behavior, acts, which create culture and not ideology and knowledge (theory of commitment). The culture of risk therefore only develops with a "practice" of risk.

The exercises rarely, if ever, involve the inhabitants. A few exemplary systems aim to include risk in the daily lives of residents: let us mention the Rhône plan and its artistic component (depicting the flood level by painting tree trunks, interactive theater play, etc.) or the CIT'IN CRISE[12] game, where for two hours, citizens learn good practices to prepare for evacuation plans, but in a playful way. Each player takes on the role of a personality (mayor, school principal, etc.) and reacts according to a well-defined scenario. Two crisis cells are set up. One has the mission of managing the crisis situation and the other cell will have the objective of confronting the players with events and hazards in order to test their reactivity. They have a model with game supports (cards, explanatory cards, tokens) and are supervised by facilitators. Then comes the debriefing to put it into practice at the local level. Let us also mention an interactive GIS, ZIP-ZICH, which allows us to visualize several flood scenarios in Ile de France. However, the effectiveness of purely virtual tools is questionable, especially for young people, who are used to video games and for whom the representation of impacts could lose its realism.

The involvement of the population can be done through mobilization days on "life-saving" exercises in case of crisis. In New Zealand, the ShakeOut Day brings together the population to learn what to do in the event of an earthquake or tsunami. In 2020, the October 15 event mobilized 680,000 people (15% of the national population) who registered via the "Getready" platform[13]. Training also involves students. In Japan, schools systematically train students in life-saving techniques and drills to learn the right actions in the event of terrorist attacks, which have become widespread in France since the 2015 attacks.

---

12. See: https://citincrise.mines-stetienne.fr/notre-offre/accueillir-citin-crise/.

13. See: https://getready.govt.nz/involved/shakeout/share-the-shakeout/.

2.4.2.2. *The integration of citizen initiatives in crisis management*

One of the challenges of crisis management is the integration of citizen initiatives in crisis management. The emergence of social media has disrupted the functioning of institutional and traditional actors in crisis management.

The potential of social media in the broad sense of the term is enormous. They formalize and multiply the participation of citizens, who are in fact the first actors of their care when they are affected by a disaster. They are a virtual place of crystallization of citizen initiatives in crisis management[14]. However, the challenge of their integration in crisis management is not solved. Security actors feel the potential of social media but are often unable to integrate them in their crisis management due to the lack of time, skills or methods. These citizen initiatives take many forms, from the creation of dedicated web pages, to participatory mapping made available to emergency services, to help in the search for missing persons.

2.4.2.3. *Anticipating behavior: knowledge of the mental representations of crisis management actors*

Crisis management is affected by myths about the perception and mental representations of the different actors. Each actor (authorities, population, emergency services, etc.) acts according to representations of the risk, which are sometimes erroneous, as well as and above all according to the perception of the other actors.

One of the most widespread myths is that looting is systematic after every natural disaster. Of course, looting exists and maintaining order is an inescapable task during crisis and post-crisis. We saw this during Hurricane Irma in Saint Martin in 2017. But this type of behavior is at the margins. Surveys in Germany (Lorenz et al. 2018) show that 42% of people think looting may occur in a disaster, but a concurrent survey of security forces attests that looting is very rare.

According to Quarantelli (1992), there are no more uncivil or unreasonable behaviors in times of crisis than in times of no crisis[15]. People have their reasons. The fact of not responding to official injunctions is not proof of "unconscious" behavior. We might call the behavior of a person who instead of moving away from danger, moves closer to it "irrational", which could be called "risky behavior": for example, returning to a flooded building while the water is still rising or refusing to evacuate.

---

14. See: https://theconversation.com/gestion-de-crise-mieux-integrer-la-reponse-des-citoyens-141741.

15. See this interview by Quarantelli (1992): https://youtu.be/E2HelEkQq2E.

Other myths pollute the relations between actors during crisis management. The myth of popular panic, that is, the belief that a certain message will trigger panic in the population, is also widespread among political leaders or emergency services. Lorenz et al. (2018) point out that rescue forces train for panic, even though panic is quite rare by the admission of these security and rescue forces.

The rumors that run through the opinion are obviously an antiphon of crisis management, and there are many examples of the dissemination of false information, especially on a cover-up of human tolls by the authorities (Moatty et al. 2019). Again, it should be emphasized that social media and the Internet have not created this type of interference even though they may have amplified its effects (Xiao 2018). Another fantasy is the fear of epidemics following natural disasters. Epidemics directly related to natural disasters are rare, and the famous example of cholera that affected Haiti after the 2010 earthquake shows that the contribution of these epidemics is often exogenous.

The image of a passive population, "shocked" by the disaster, must also be put into perspective. Dynes (1994) shows that situational altruism is the dominant behavior in a crisis and that "the majority of relief and immediate assistance is provided by survivors, not by emergency services" (Lorenz et al. 2018, p. 359).

Planning crisis management based on presuppositions, prejudices without knowing the ways in which the actors and populations concerned operate, their degree of trust in the authorities, the social and economic capacities to help each other, etc., exposes major dysfunctions (Drabeck 2010; Weiss et al. 2018).

2.4.3. *The importance of logistics*

During a crisis, the need for expert skills evolves. From "hazard" skills, we move towards a need for assistance to people and logistics. Expertise is often concentrated on hazards, whether for epidemics, floods, etc. However, the management of large-scale crises requires logistics that are often neglected in favor of immediate assistance to people (earthquake). The lack of logistical means, delivery problems, the adequacy of means to needs are recurrent. During hurricane Irma in the West Indies in 2017, public authorities lacked the means, particularly nautical means, to reach the affected northern islands. However, the owners of inter-island launches or passenger ships affirm that they were not solicited.

The management of long and/or large-scale crises requires skills in terms of logistics. During major crises accompanied by massive destruction (earthquakes, hurricanes, conflicts, etc.), the complete disorganization of networks requires the re-establishment of a logistics chain. During the earthquake of January 12, 2010 in Haiti, the port of Port-au-Prince and the roads became impassable. The airport operated in degraded mode for two months, and a large part of the aid was transported from the Dominican Republic by road. These logistical difficulties were compounded by the poor state of the road network before the disaster. In addition, crisis management was complicated by conflicts between actors, particularly between the U.S. military and the United Nations, over control of logistical resources. The skills of logisticians are therefore much sought after by NGOs, which invested in this key issue earlier than the states. The question of the "last mile", that is, the final accessibility to victims in case of a destructive event (earthquake, conflict, massive attack) is crucial.

These logistical problems are sometimes linked to malfunctions or to the lack of public/private partnerships. Crisis preparation is costly. If states want to size up their resources to deal with a major crisis by anticipating all types of crises (health, technological, natural), this generates exorbitant costs in terms of human and material resources, resources that by definition are only useful every 20 or 30 years. If it is admitted that the states cannot satisfy all the needs in case of a major crisis, which is understandable, this must be compensated by a pooling of means (private, international, etc.) and prior agreements between private actors and public managers of the crisis. These agreements are conditioned by the good state of relations between these actors before the crisis.

2.4.4. *Population management: the issue of evacuation*

In the panoply of crisis management tools, a special mention must be made of evacuation, an eminently territorial crisis management method. The evacuation of people, animals or exposed goods (evacuation of works of art from the Louvre in the event of flooding of the Seine) is undoubtedly the most delicate decision, as its implementation can be difficult and the undesirable effects numerous. Preventive evacuation, that is, within a fairly long period of time before the effects of the hazard become apparent, is possible for certain risks. It is not always possible, especially for sudden phenomena without warning signs (airplane accidents, explosions, terrorist attacks, earthquakes, etc.). It is possible when the mounting hazard is gradual and spatially circumscribed with more or less precision. Here again, crisis management engineering provides many technical references on this type of exercise: links with forecasting, time frames, spatial management, ordering party, evacuation modalities from the safety on site (vertical evacuation in the floors in case of marine submersion, for example), constrained or free mode, number of people involved, refuge areas, etc. (Peroche 2016; Piatyszeck et al. 2018b).

Above the technical aspects, the success of evacuations is linked to the preparation of the populations and to the representations that the operation covers in people's minds. The relationship with the territory, the confidence in those giving the orders, the feeling of safety in abandoned places and places of refuge are all parameters that determine the success of an evacuation. When asked about their refusal to evacuate during the Xynthia storm in western France in 2010, people replied that they had not backed down in the face of Hitler and that they would not back down in the face of a flood. Evacuation is experienced as a retreat, a defeat, an abandonment of place. The relationship with the territory is different in the United States, for example, where hurricane warnings trigger spontaneous evacuations that do not fail to pose problems in other respects (traffic jams, gasoline shortages).

Some places are particularly difficult to evacuate, such as health-care facilities and, for different reasons, prisons. For the former, the evacuation of patients or residents of retirement homes requires a lot of personnel and disturbs already fragile organisms. Studies have shown excess mortality in evacuated retirement homes during the June 2010 floods in the Var department (Mantey et al. 2012). For this type of establishment, we should look for a location outside the risk zone or, failing that, a local security system (with a floor outside the risk zone that functions independently).

It should also be noted that there are many contradictions in the instructions that may be given in the event of a multi-hazard crisis. In the case of a chemical incident, containment may be recommended, but if the incident is caused by a tsunami or a flood, the containment instruction may not be relevant. The same is true for evacuations related to storm surges.

2.5. Conclusion

Crisis management by definition is the response to residual risk. Preparing for crisis management therefore presents a paradox. On the one hand, it requires a very strong link with prevention, as these prevention plans must include measures to be taken in the event of a crisis and allow the territory to be developed in such a way as to better manage the crisis and allow for rapid recovery. However, this link is not obvious at all because it requires accepting that prevention has limits, including therefore preparing for the crisis using other scenarios than those used for prevention. Yet, focused on "emergency" expertise, crisis management is often focused on the hazard or the danger, while approaches centered on assets at risk have difficulties to be addressed in operational planning. Crisis preparation must move away from prevention as the reference horizon of crisis. It is necessary to prepare for what is outside the prevention schemes.

There are many tensions in crisis management. The first one is between the sectorial specialization of the actors in the crisis management chain and the need for a holistic vision of crisis management. At the same time, since the beginning of the century, we have witnessed a tension between this growing expertise and the irruption of the layman via the media (including social media). The advent of "open science" and the exponential dissemination of information pose a major challenge to crisis management actors. The challenges of crisis management are numerous. They are largely based on building mutual trust between institutional and private actors, between managers and populations, between crisis management professionals and populations, between experts and politicians. Crisis management planning tools have made great progress, but the psycho-sociological, political and ethical obstacles are still numerous. One of the prerequisites for building this trust is the sharing of a diagnosis on the potential impacts of a crisis. This confidence building is cultivated through joint learning that moves crisis management from plan to safety device, and then to action.

3

Learning from Crisis Management

Freddy VINET[1], Alain CHEVALLIER[1], Hoilid LAMSSALAK[2] and Dimitri LAPIERRE[2, 3]

[1] *Université de Montpellier III, France*

[2] *RisCrises, Alès, France*

[3] *CERISC, Aix-en-Provence, France*

The increasing complexity of crises, the emergence of crisis management engineering, the growing number of actors and the multiplication of tools (warning, planning, etc.) make it more and more imperative for managers themselves as well as for exposed populations to learn crisis management. Crisis management cannot be improvised, it must be learned. The challenge is to move from empirical know-how based on experience to crisis management engineering based on the capitalization of learning, the acquisition of "reflexes" and the use of generalizable and transferable tools and processes. Indeed, going through a crisis and the experience drawn from it is the first form of learning. However, given the mobility of populations and the low frequency of catastrophic events, capitalizing on experience based solely on what has been lived is not likely to give rise to sufficient preparation.

The training of crisis managers does not stop at the planned processes. There is even a tendency to say that learning to manage a crisis begins where planning ends and that one must be able to make decisions in spite of uncertainties when the course

Territorial Crisis Management,
coordinated by Richard LAGANIER and Yyvette VEYRET. © ISTE Ltd 2022.

of the crisis does not follow the plans initially made. Today, it seems necessary to prepare crisis decision-makers to deal with the unexpected (Lagadec 2012).

Crisis management is based on invariant, intangible and permanent methodological tools and on permanent and varied training, exercises and simulations, followed by systematic feedback. These invariants are the pillars of a resilience capacity in the face of all types of crisis:

– a shared response (immediate reporting, upward and downward status updates, a log of past and future events);

– a crisis management tool (e.g. a future monitoring or summary table showing the objectives, priorities, choices and decisions of the decision-maker, distribution lists, a crisis directory that includes an increasing number of actors);

– the acquisition of a common language and the establishment of consistency between the different actors of crisis management;

– a graduated response (based on the principle of subsidiarity);

– the acquisition of automatisms and reflexes that save time in crisis management;

– an identified command structure: the decision-maker should not be confused with the crisis pilot. The decision-maker can be compared to a composer. They compose the music and decide on the tempo, the cadence and the rhythm, whereas the pilot can be compared to a conductor. The former makes decisions, the latter implements them;

– planning based on reflex cards and measures that must respond to the widest possible range of crisis situations;

– exercises and simulations;

– continuous feedback.

This implies, at the territorial level, that all of these invariants are known at least by the territorial actors. The distribution of guides, concepts, plans and standards is not enough. Only an inter-knowledge of the actors involved in crisis management and their operating methods can guarantee successful crisis management (Tena-Chollet 2012, p. 54). This inter-knowledge is acquired during numerous and varied exercises. However, this is sometimes difficult given the organization of crisis cells within the territories.

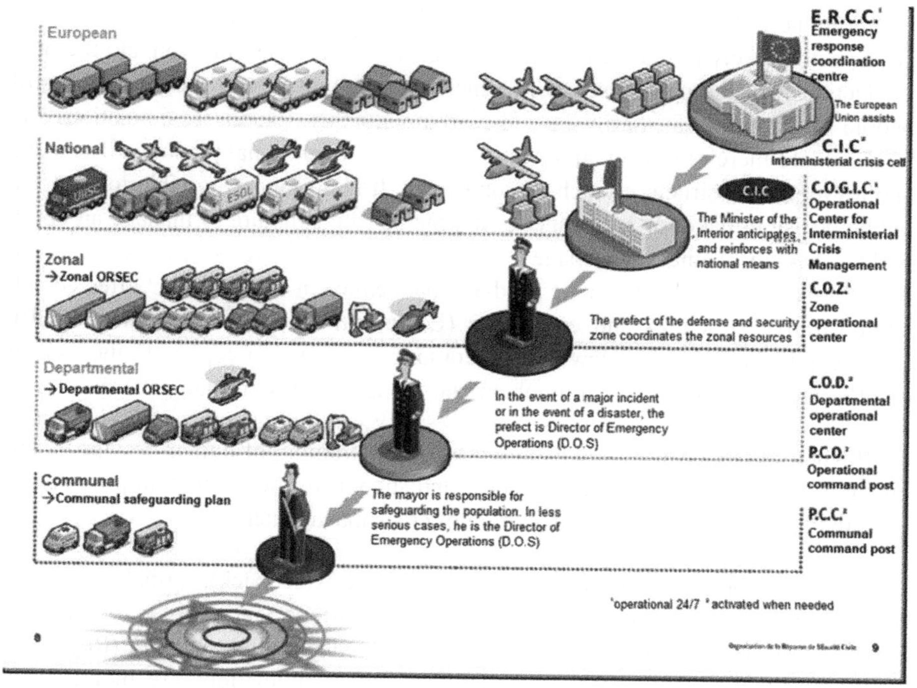

Figure 3.1. *Decision chain and operational chain in crisis management. For a color version of this figure, see www.iste.co.uk/laganier/territorial.zip*

The organization chart illustrating the management of civil protection resources in France shows five levels from the local command post to the European level of the Emergency Response Coordination Centre (Figure 3.1). However, only the mayor and the departmental prefect have decision-making powers. At the governmental level, an interministerial crisis center (*Centre interministériel de crise*, CIC) manages a crisis of national scope. The other levels are only coordination levels (COZ[1], COGIC[2]). In operation 24 hours a day, they have no decision-making power but provide their expertise to the prefects of the departments in anticipating the commitment of resources. There is therefore a dichotomy between the political (decision) and the technical (crisis management). The decision-making levels within the territories do not have a crisis room that is in operation 24/7. They have to be activated (communal command posts for town halls and operational defense centers for prefectures) by bringing in resource persons who do not meet on a daily basis,

1. *Centre opérationnel de zone*, Area Operations Center.

2. *Centre opérationnel de gestion interministérielle des crises*, Interdepartmental Crisis Management Operational Centre.

except in the field. All of this explains the difficulty and the necessity to acculturate all the actors likely to intervene in the management of a crisis situation and the little interaction with the private sector in general. Crisis units are ad hoc gatherings of people who are not necessarily used to working together. Exercises, simulations and feedback are therefore essential to test and optimize operational planning. The same applies to private structures such as companies. It is no longer possible today for a public decision-maker or a manager of a large private structure to manage an entity without having planned to manage a potential crisis, or better still to anticipate it. The crisis generated by the Covid-19 pandemic in 2020–2021 affected all administrative structures and companies (Ecole nationale d'administration 2020). Many companies had to review their operations in a hurry to cope with the lack of employees, the fall in customer numbers, the change in consumption patterns, etc.

Beyond the procedures and planning tools, the training of human resources, the preparation of leaders, and the awareness and support of the population provide the essential flexibility in the implementation of crisis management.

The oldest mode of learning is *retour d'expérience* (feedback), formalized or not, an inadequate French term that refers to the a posteriori analysis of a crisis or exercise to learn from it and improve crisis preparedness after an event (Rey et al. 2019). In addition to feedback, exercises and crisis simulations are moments of setting the mechanics of crisis management to music. From the "tabletop" exercise to the real exercise, in accelerated time or real time, between managers or involving the population, these simulations take variable forms.

3.1. Feedback: objectives, contents, actors

Feedback, in other words the ex-post analysis of a real or simulated catastrophic event, is an old exercise already practiced in the 19th century or even earlier. On November 14, 1854, in the middle of the Crimean War between the Russians and the Franco-English forces, a violent storm hit the Black Sea destroying 38 ships including a warship, the Henri IV. The then director of the Paris Astronomical Observatory, Urbain Le Verrier, led the investigation by collecting more than 250 documents from his European colleagues attesting to the circulation of the storm from west to east of Europe in the previous days. On February 16, 1855, Le Verrier submitted his report to Emperor Napoleon III in which he recommended the creation of a meteorological observation network to better predict such phenomena. Learning from a crisis in order to better anticipate the next one is the essential credo of the feedback that has multiplied and been systematized since then. While the objectives have not changed much, the procedures and forms of feedback have evolved and diversified.

3.1.1. *Objectives of feedback*

Feedback is defined by the World Health Organization as

> After action reviews and simulation exercises are voluntary instruments that help countries assess their operational capability for public health preparedness and response. They provide functional assessments, and play a key role in identifying strengths and gaps in the status and implementation of IHR capacities. They can be used to review, validate or 'stress test' the capacities found/reported by other IHR MEF instruments – for example, by looking at how effective a program or system is versus the extent to which relevant policies are in place. Used by many organizations and across sectors, AARs and simulation exercises are important learning tools and effective methods for informing stakeholders and identifying best practices, challenges and key lessons that can help improve response capability. Both tools can help identify the root causes of preparedness gaps that, if addressed, can improve future responses to health emergencies (WHO 2018).

For Patrick Lagadec,

> [...] feedback consists of drawing lessons from a concrete experience of crisis management for other situations of strong turbulence. The quest for lessons concerns a much broader field than the mere technical resolution of a tool or process failure. It is a question of questioning, in particular, individual and especially organizational reactions to an acute ordeal, which may have destabilized the "system" (company or groups of entities involved) (Lagadec 1992).

Analysis of a crisis and its causes is the primary objective of feedback. Feedback looks for the immediate and deep (structural) causes and examines the evolution of the territories over the long term. Indeed, it is impossible to understand a disaster without reconstructing the vulnerability trajectories of the affected territories (Vinet et al. 2011; Barclay et al. 2019; Duvat et al. 2021). The extent of the damage suffered during the Xynthia flooding on the French Atlantic coast cannot be understood unless the storm is placed in the long history of Atlantic coastal territories, marked by residential tourism, aging populations and uncontrolled urbanization, itself favored by agricultural recovery. The various forms of feedback point out the lack of preparation for the crisis or the unsuitability of the building (lack of floor). But they do not always go into detail, for example on the inadequacy of electric shutters which, in the event of a power cut caused by a flood, a storm or a fire, prevent people from escaping from their homes.

3.1.2. *Feedback times: feedback and territories*

Three types of feedback can be distinguished according to the time scale considered: hot feedback is practiced almost during the crisis or post-crisis. It is partial, sectoral and aims to inform on the operations to be followed (post-crisis, recovery, reconstruction). The post-crisis or classic feedback generally begins a few weeks to two or three months after the disaster and meets the objectives mentioned above. A third type, long-term feedback, evaluates the impacts of disasters over the long term, particularly the territorial impacts.

3.1.2.1. *Hot feedback*

The first assessments of impacts intended to size the post-crisis response and reconstruction can be classified as "hot" feedback. They are often "sectoral" and provisional, as it is difficult to immediately grasp the full scope of a crisis (Molinari et al. 2017). These forms of feedback based on space technology, such as the European Copernicus consortium, which sometimes provides damage[3] estimates less than 24 hours after a disaster. Unfortunately, the results of these hot forms of feedback remain engraved and the provisional assessments are often considered as definitive due to the lack of a reanalysis afterwards. This was the case when, a few days after Hurricane Irma, the Copernicus system showed that 95% of the buildings on the island of Saint Martin had been affected, which was transformed in the media into the destruction of 95% of the buildings. The reality published in a report one year later is more nuanced (Copernicus 2018).

This type of hasty assessment is also common in the human toll of disasters. During the Vargas floods and landslides in Venezuela in December 1999, an initial death toll of 30,000 was recorded in international databases[4] and feedback, attracting the sympathy of the media and NGOs worldwide. However, a specific re-evaluation of this death toll by anthropologists showed that the real death toll was in the order of several thousands (Altez et al. 2005).

For crisis management in the strict sense of the word, hot debriefing is the first step in the feedback process. It is carried out whatever the perception of the management of the event encountered: whether there were major dysfunctions or not, all management, intervention or participation in an event requires considering the purpose as well as being interested in the "why" and the "how". On-the-spot debriefing is essential because it allows the oral sharing of the experience and

3. https://emergency.copernicus.eu/mapping/#zoom=2&lat=13.54457&lon=27.49718&layers=00B0T.

4. https://public.emdat.be/data.

constitutes an essential basis for post-crisis learning. It is also during these collective sequences that participants can question themselves and obtain factual details about the situation they have just encountered (or a targeted part of it). Finally, it is also during the debriefing that a first global analysis can be made at the group level.

Today, there are many recommendations for structuring a hot debriefing (make sure that everyone expresses themselves, do not pass judgment, identify the points that worked well and the areas for improvement, etc.) and the posture of the "debriefer" appears to be one of the essential links in the success of the debriefing and the commitment to a reflexive process (this is known as "facilitating").

Hot feedback also serves as an outlet and a "defusing" process for the participants. After an event or an exercise, it is important to identify the people in the crisis unit who did not experience it well. Indeed, to avoid any traumatic shock, even the slightest, it is essential that people express their feelings, their experience of the event. In crisis situations such as the Covid-19 crisis experienced by hospital staff, the decisions taken, sometimes extremely harsh, are considered personal failures. This accumulated stress must be reduced at all costs through internal feedback. Internal debriefing by supervisors or peers can induce a form of self-censorship. The practice of questioning is preferable to a raw restitution of observation elements.

3.1.2.2. *Post-crisis feedback*

The understanding of territorial crises requires broader, multisectoral and multi-temporal, integrated feedback. "Integrated" feedback is carried out by public bodies, non-governmental organizations or research consortiums that seek to have a holistic vision of the crisis in its territorial, social, economic and legal components. They include a number of themes from the identification of impacts to reconstruction and crisis management (Huet 2005; Moatty 2015). The methods used and the data collected depend on the type of sponsor and the means implemented (Table 3.1).

Post-crisis feedback, when conducted or commissioned by institutional actors, is political in method and content. The choice of the people heard must respect a balance between all the actors concerned. Feedback is therefore also a forum where the various stakeholders in a given subject (technological accident, health crisis, etc.) must be able to express their point of view. Although the diagnosis is intended to be as "objective" as possible, the interpretations and recommendations drawn up are not generally neutral. They reflect the main ideas of a prevention policy for the risk in question and guide the choices of those responsible. The choice of a particular option to reduce a particular risk may lead to new legislation or new measures. Feedback from the floods of 1999 and 2002 in the south of France and from the explosion at AZF in 2001 provided the framework for the development of the 2003[5] "risks" law. Feedback in this sense feeds action and serves as a lever for territorial planning and management.

1. Preparation	
Step 1.1	Determine the objectives and the spatial and temporal scope of the feedback
Step 1.2	Identify the partners associated with the feedback
Step 1.3	Determine the level of feedback
Step 1.4	Determine the format of the feedback
Step 1.5	Develop the implementation schedule
Step 1.6	Designate the pilot and write a summary note to the sponsor
Step 1.7	Inform the contributors and specify the organization of the feedback
2. Implementation	
Step 2.1	Ensure the collection of information
Step 2.2	Verify and synthesize the information
Step 2.3	Develop and test recommendations
Step 2.4	Restore the elements by developing a report, a synthesis document and various valorization media (atlas, website, etc.)
3. Verification, follow-up and valorization of the feedback	
Step 3.1	Disseminate the report to all contributors and to the sponsor
Step 3.2	Organize a plenary meeting and thematic feedback meetings
Step 3.3	Make the feedback available to all the actors concerned, the media and the population

Table 3.1. *Objectives and feedback content (source: Ministère des Solidarités et de la Santé (2019, p.19))*

3.1.2.3. *Long-term feedback and real-time monitoring*

However, "classic" feedback does not cover all subjects. It does not take into account the long-term impacts of a disaster. The purpose of long-term feedback is to study the evolution of a territory several years after a disaster. The questions are multiple (Leone 2007; Moatty 2015). What were the structural impacts of a crisis on the territory? Did the territory evolve towards greater resilience? Was the post-crisis period conducive to adaptation to risk? This type of feedback helps to reframe and put into perspective prevention or reconstruction policies, which are often decided in

5. Law no. 2003-699 of July 30, 2003 on the prevention of technological and natural risks and the repair of damage.

haste and in reaction to the disaster. However, it is rarely practiced except by research teams who return to the areas concerned, five, 10 or 20 years later. One example is the research of d'Ercole which, in a 1996 article, looks back at the earthquake that largely destroyed the city of Popayan in Colombia (d'Ercole 1996). The lack of interest of public authorities in this kind of return to the past is flagrant. Yet, it is the richest exercise for understanding territorial dynamics in the face of crises.

Another form of feedback is the real-time (*in itinere*), continuous monitoring of the territory affected by the crisis in the years following the disaster (see the results of the Tirex research project on the territories affected by Hurricane Irma in the Antilles) (Tirex 2021). The other difficulty with this type of exercise, in addition to the lack of interest on the part of public authorities in such a distant return to a disaster, is to determine the real weight of the disaster in relation to the structural and external evolutions of the territory (Bankoff et al. 2004). Where does the geography of development begin? Where does the geography of risks and disasters end?

3.1.3. *Feedback methods and contents*

3.1.3.1. *Internal feedback*

Feedback is characterized by its degree of dissemination and openness. A distinction is made between internal forms of feedback, for example, conducted by a player (private or public) and intended to test and improve the player's internal systems. Examples include the internal feedback of Enedis[6] after storms or the feedback of organizations in charge of rescue operations. These forms of feedback are sectoral and only envisage a partial vision of the crisis. They are generally not made public and may include a search for responsibility that may lead to legal consequences. This judicial aspect is a hindrance to the dissemination and opening of the feedback to other actors.

The level of feedback chosen depends on whether the client is a sectoral player in crisis management (motorway concessionaire, insurer, private company) or a player with territorial and generalist competence (State, ministry, parliamentary committee, etc.). In the political domain, we note the proliferation of parliamentary inquiries by commissions of federal or national elected officials over the last 20 years. This type of "institutional" feedback consists mainly of gathering the views of

6. *Service national français de distribution de l'électricité,* French National Electricity Distribution Service.

the main actors and victims of the crisis and of various external experts. They proceed by iterative interviews until saturation, that is, when the new interviews no longer reveal any new information and the information gathered becomes redundant (without necessarily being concordant). The content of the feedback will depend on the extent of the phenomenon, the emotion or controversy it has aroused in the population and also the evaluative culture of the participants.

3.1.3.2. *Open feedback*

The increasingly wide and easy dissemination of information via the media and social networks calls for open feedback, that is, in which all those involved, including the victims and the general population, can follow and contribute to the investigation. Let us take the example of the British approach following the fire at the oil depot in Buncefield, north of London[7], on December 11, 2005. The British authorities set up a dedicated interactive website, open to all, where all the information related to the investigations was put online, while reminding us that some of this information was still being analyzed.

Thus, Taf Powell, the "Buncefield Investigation Manager", expresses himself, from the "progress report", that he was asked by the Major Investigation Board to prepare for the second meeting a progress report for the Health and Safety Executive and the Environment Agency. The purpose was to retrieve information and track the progress of the investigation and any events established on this issue[8].

All of these elements and reports have been posted on a dedicated[9] government website. The initial report and the three progress reports were prepared by an independent team.

3.1.4. *Feedback: a tool for sharing and disseminating learning*

Dissemination of feedback is essential to share the knowledge gained. Too often, for fear of the legal consequences or political backlash that may result from disclosing what has not worked, feedback is buried. This undermines knowledge

7. https://www.hse.gov.uk/comah/buncefield/index.htm.

8. "At the first meeting of the Major Incident Investigation Board (MIIB) on 24 January I was asked to prepare, for the second (10 February) meeting, a progress report on the Buncefield Major Incident Investigation by the Health and Safety Executive (HSE) and the Environment Agency (EA). This was to bring together relevant background material and to cover progress with the investigation and such facts as have been established to this point."

9. https://www.gov.uk/government/publications/the-buncefield-investigation-the-government-and-competent-authoritys-response-november-2008.

sharing, even when unsourced directives propose changes to the procedures in place. In the feedback section of the French Department of the Interior[10], there are only two documents online in the homeland security section, which raises questions.

The purpose of feedback is not only to identify good practices to improve after a crisis, but it also aims to prevent future crises. To do this, it must not be limited to crises experienced in the territory, because postaccident analysis is not limited to what has been experienced, but must also analyze near-accidents and near-crises.

Figure 3.2. *Fire at an oil depot in Buncefield, North London (realization: René Dosne). For a color version of this figure, see www.iste.co.uk/laganier/territorial.zip*

Thus, it is important to also analyze the crises suffered by other territories, including abroad, with similar characteristics and to model them on our own crisis management systems in order to assess their robustness. Following the Buncefield

10. https://www.interieur.gouv.fr/Le-ministere/Securite-civile/Documentation-technique/Les-retex.

fire mentioned above, a mission of the Paris fire department – in partnership with a French oil company – was to determine whether this accident could occur in French oil depots. An effort had been made to graphically popularize the potential impacts of an explosion on an oil depot. The answer after investigation was negative because the French regulations are more draconian than the British ones, and several elements could not have occurred in France. Feedback is not only curative but also preventive. We have to look at events that have happened elsewhere and see if our rules and procedures would have made it possible to avoid a crisis or at least reduce its severity.

Feedback is an essential tool for communicating information and participating in the creation of what is commonly called a "risk culture". They take advantage of both the freshness of the event and the external view of the territory and the risks that they face.

If there is progress to be made, it is in the valorization and dissemination of feedback. Often, resources are allocated to the feedback itself while neglecting the dissemination work. This dissemination work requires specific skills, time and energy, often more than is needed to establish the feedback itself.

After Hurricane Irma hit the Caribbean in 2017, a research consortium conducted integrated feedback on the northern islands (Saint Martin and Saint Barthélemy)[11]. A large part of the project was dedicated to the valorization of the results of the feedback. Based on an understanding of the cultural, societal and political-administrative contexts in which the lessons learned from the feedback were transmitted, the aim was to evaluate individual and collective expectations in the post-crisis period. The tools for transferring knowledge from the feedback took several forms (serious game, geo-visualization platform, preventive reconstruction guide, indicators) and were tested on site with local actors and populations. The dynamic and interactive geo-visualization tools (web mapping, storymaps, chronologies) allowed us to visualize the impacts of the cyclone and the reconstruction[12] phases. The information acquired served as the basis for the construction of multiple and graduated crisis scenarios. The development of a guide for a territorial and preventive reconstruction responds to a need expressed by the actors.

11. https://tirex.univ-montp3.fr/index2.html.

12. See the storymap on crisis management: https://storymaps.arcgis.com/stories/6cb36221e8b743f3b89859afa77a8095.

3.1.5. *Conclusions on feedback*

The scope and methods of territorial crisis feedback have expanded considerably over the past 20 years due to the triple effect of the complexity of these crises, technological progress (mapping, telecommunications, digital technology, etc.) and the demand for transparency on the part of the actors and populations exposed.

The evolution towards greater transparency is imperative if we do not want to fuel suspicion towards the authorities or the large private actors deemed responsible by default for disasters. New information and communication technologies have multiplied the possibilities of publicizing and disseminating feedback. However, certain limits remain. The first issue is the degree of detail and technicality of the feedback in relation to its degree of publicity. This leads to tensions between the need for comprehensive and synthetic feedback that requires ever higher levels of technicality and expertise.

Publicity and the growing need for transparency require balanced feedback that is not only "incriminating", what Jean-Luc Wybo once called "positive" feedback. The implementation of contradictory and public feedback should undoubtedly be developed. Another issue is the sharing and appropriation of the conclusions of the feedback and the integration of "good practices" into the "current" management of the territories.

3.2. Crisis management exercises and simulation

The types of exercises are numerous and form a spectrum that unfolds according to their degree of realism and complexity. An exercise in situ, on the real terrain where a crisis and its management can occur, in real time and with all the participants concerned, including the population, will obviously be much more favorable to immersion and awareness. However, this complexity is not to be sought systematically. It depends on the objectives assigned to the exercise.

3.2.1. *Objectives and prerequisites*

Where training courses can often be marked by a lack of motivation on the part of the trainees, crisis management simulations offer them the opportunity to engage in a dynamic behavioral process. The actors are then immersed in an environment that is at the same time immersive, contextualized and realistic. These are the first essential building blocks to sensitize and train actors to act in an emergency, in the immediate, under controlled stress. Indeed, due to the suddenness of a crisis, the strategic management of a crisis is particularly complex and crucial, especially in the first moments. This necessity is all the more justified since the crisis unit is an ephemeral crisis team that operates and shuts down according to the pace of the crises (Dautun et al. 2013): this ephemeral character constitutes a real obstacle to the management organization because it prevents its members from adopting work reflexes.

In a global way, the realization of a crisis management simulation must answer a certain number of questions in order to establish its pedagogical approach: the type of exercise and the definition of space–time parameters.

The definition of the exercise objectives is crucial for its parameterization (Table 3.2). Is it to meet a regulatory[13] requirement? Is it an awareness exercise, a training exercise for laypeople with little crisis management experience, or training for professionals? Is the exercise a single exercise or part of a multi-faceted and progressive training program?

Regardless of the type of exercise conducted, trainers must identify the associated educational objectives, ensure that these objectives can be evaluated by the training mechanisms used, be able to evaluate them with criteria defined upstream, as well as be able to report back to the participants (Lapierre 2016). This work must be consistent with the organization of the different types of crisis exercises.

In addition to simulation exercises, there has been a desire in recent years to diversify the ways in which crisis management is approached. The first transformations are visible in the variation of learning approaches: the contributions in terms of playfulness, in particular, tend toward training through games (gamification). More and more trainers are interested in games (escape games) as a new training lever: the concept of leaving participants with total autonomy, in a closed environment, in search of clues and avenues of reflection to solve problems encountered in a limited period of time, is attractive. The concept, which can be transposed to the themes of crisis management, proves to be immersive while raising awareness: CrisiScape, created by the *Institut Mines Télécom – IMT Mines Alès*, is a

13. While the regulatory obligation constitutes, we agree, minimal motivation, it sometimes has the virtue of obliging the actors to train, for example, for exercises before each cyclone season or in the industrial field, exercises on the internal operations plans for the company or particular safety plans in schools.

first step in raising awareness. Applied to the problems encountered by a crisis unit, CrisiScape has had conclusive[14] initial feedback from users. Immersion in virtual reality as a learning tool is not new (Mitchell 1997). The investigation of these new learning avenues is being extended through the creation of other workshops: the reflections carried out around *Crise à la Carte* give us a glimpse of a new way of carrying out tabletop exercises, which are often associated with an asynchronous character and therefore demotivating for the participants. By proposing a playing card mechanism to facilitate the handling, understanding and use of an emergency management document, the development of skills can be achieved through serious play.

3.2.2. *Organizing a crisis management exercise: methodology*

When organizing an exercise, there are a number of different issues that form the inputs needed to develop the exercise content (Table 3.2). The elements of exercise design are a compromise between human and material costs and organizational complexity, on the one hand, and realism, on the other hand. An in situ exercise in real conditions, particularly concerning the timeline involving all the stakeholders including the population, will require significant technical, financial and human resources. This deployment of resources is rarely justified in view of the benefits derived and the risk of "failure" which can have a counter-productive effect. If the exercise "fails" or demonstrates major organizational flaws, this can increase mistrust of the population.

Tabletop exercises with a limited number of actors and in accelerated time are obviously easier to organize, but they only give a biased view of the conditions encountered in a real crisis.

The choice of participants is obviously crucial to disseminate learning in the social fabric. The addition of actors with little experience in crisis management but with knowledge of the risk is often decisive. One example is the EU Richter Caribbean 2017 exercise during which social landlords in the French West Indies were at the heart of a seismic and tsunami crisis management system.

Second, organizing an exercise requires clear characterization of the three phases of the implementation process (Lapierre et al. 2020).

14. Crisis management training redesigned by the LSR team at IMT Mines Alès: https://www.innovation-pedagogique.fr/article7286.html.

Parameter	Terms and conditions	Observations
Educational objectives	1- Those related to the conduct of the strategic crisis response 2- Those related to the organizational aspects of the crisis unit 3- Those related to the conduct of strategic crisis communication	An exercise without objectives is of no interest and their choice conditions the construction of the exercise. First, it is necessary to identify the different components to be evaluated from the three main themes related to strategic crisis management. Each of them is associated with a set of educational objectives with specific evaluation or observation methods. Generally, due to lack of resources, the evaluation of educational objectives relies on the experience of trainers or observers or, orally, during debriefing discussions. The deployment of a complete evaluation system requires strong mobilization of resources.
Location	1- Indoor (tabletop) 2- In a simulator (equipped room) 3- In situ	
Participants	1- Experts 2- Experts: laypeople 3- Populations	A distinction is made between crisis management professionals (firefighters, etc.) and expert-professionals with knowledge of the risk or of the territory and the populations.
Temporality	1- Actual time or 2- “Accelerated” time	
Immersion	1- Scenario (by telephone solicitation) 2- Scenario (by radio solicitation) 3- Plastrons on the ground 4- 3D modeling of the current event 5- Social network simulation tools	An exercise scenario can be run through a single channel if necessary, the telephone. To increase immersion, immersive levers can be activated and deployed, but they require more resources.
Field	1- Complete crisis 2- Partial	The exercise concerns the complete crisis or a part of it (alerts, evacuation, etc.)
Risk	1- One risk 2- Multi-risk	The exercise may include domino effects (Natech, NatSan).
Budget	Resources available (human, technical and financial)	The “budget” parameter obviously depends on the other parameters, that is, the scope and complexity of the exercise.

Table 3.2. *Elements of crisis management exercise parameterizations. Modality ratings range from 1 (inexpensive and uncomplicated) to 5 (expensive and complex to organize) based on several sources including Chevallier (2009) and Tena-Chollet (2012), modified*

The first phase is dedicated to the preparation of the exercise. It takes place several weeks to several months before execution of the exercise, depending on the scale of the exercise and the mobilization of the players involved. It is a question of determining the needs of the exercise and the trainees in order to translate them into the educational objectives of the exercise. The exercise scenario must include elements that promote both the immersion of all participants and the credibility of the situation encountered. Observation and evaluation tools are used to evaluate the educational objectives but not solely: they must be constructed in order to understand how the crisis unit was organized and functioned, and how it made decisions.

The second phase is the execution of the exercise. To carry out the exercise scenario, the trainers rely on a team of facilitators in charge of injecting the different elements of the scenario (such as major events, panicked calls from residents, requests for specific information and requests for action to be taken). Observers present with the participants during the exercise have a double mission: the first is to fill in their evaluation tool created before the exercise, according to the chosen objectives; the second is to be a relay for facilitators. Since the facilitation team is often "blind" (i.e. physically distant from the participants), observers can communicate with them in order to point out possible drifts among the trainees, unintentional events or technical problems. One of the techniques that allows the facilitator to follow the events is to consider that the decision-maker is not present, which avoids putting too much pressure on the players. The decision-maker remains at the facilitation level and can call or be called for regular updates from the players, thus allowing to refocus the thread if needed.

The third and final phase involves analyzing the exercise and reporting the results. To achieve this, there are several steps: first, it involves using the data entered into the observation and evaluation tools during the exercise. Second, a debriefing takes place a few moments after the end of the exercise: this allows all the participants to discuss the experience of the exercise, the points for improvement (not negative ones, as it is always necessary to be positive), the difficulties encountered and the first areas for improvement. Beyond the shared experience, it is important that everyone identifies the elements to be retained from the exercise (Fanning and Gaba 2007). A few weeks later, the organization of a cold debriefing is proposed with the same actors: it is essentially during this new meeting that the pedagogical expectations are addressed and contextualized in relation to the scenario played out. It is at the end of these two debriefings that a document dedicated to the formalization of the feedback is created, including all the elements to be retained, the understanding and the origin of the failures and dysfunctions, as well as all the recommendations to be taken into account in order to improve crisis management organization.

3.2.3. *Types of exercise*

We frequently encounter three types of exercises, presented here from the most to the least onerous in terms of organization, with different objectives:

– The first one, called "full-scale exercise", mobilizes all the actors concerned at the strategic, tactical and operational levels. The full-scale or "real" exercise, in the sense that it is supposed to reproduce the temporal and geographical conditions of a potential accident, is particularly interesting for evaluating the coordination between actors, the operationality of procedures and the inter-service communication mechanisms. However, it is rarely applicable because of the costs (human and financial) that it generates. Moreover, this type of exercise is often publicized in the media to show the state's capabilities in the face of a crisis. The problem is that the exercise loses its spontaneity and quickly becomes a demonstration with issues such as the location of the command post which must be compatible with the presence of the media and one or more ministers (Chevallier 2011). What can we say about an "attack exercise" that took place, as in 2008, in the inner courtyard of the Paris fire department headquarters, a highly unlikely place for this type of event? However, the purpose of these exercises is to legitimize the work of the field personnel.

– The second type of exercise is called "functional": its approach, based on action and reflection, allows a global approach to the situation encountered. The latter is simulated with the help of a scenario that will put the crisis unit in a dynamic and above all experiential situation. The notion of experiential learning (Kosarzicki et al. 2005) takes on its full meaning here: in an environment where the repercussions of their decisions will have no real consequences, the participants can learn about both "taskwork" (from the implementation of standard procedures to creative decision-making) and "teamwork" (all the processes specific to the organization and functioning of the group). This type of exercise is sometimes called "simulation", to which we will return later.

– Finally, the third type concerns the "tabletop" or indoor exercise. A simulated event generates discussions between the participants on the reflections to be carried out, the actions to be taken with the aim of solving the problem encountered. While this exercise is interesting because of its reflective approach and the low costs involved, it suffers from a lack of immersion, realism and dynamics which limits its pedagogical scope compared to a so-called "functional" exercise (Tena-Chollet 2012).

A good definition of the space–time parameters of the exercise brings a minimum of realism. If crisis management simulations have progressively become an essential tool to train crisis decision-makers, it is above all because the learning of its participants is at the heart of a training session. The opportunity to learn or improve by practicing skills remains one of the key points of a simulation.

To achieve this, it is important to consider different prerequisites directly related to the exercise scenario, namely its credibility, its pedagogical character as well as its interactivity (Fréalle 2018). Indeed, the credibility of the scenario favors the engagement of the trainees in the simulation while its pedagogy is linked to the need to tend toward the acquisition or improvement of knowledge, for example. Moreover, the notion of interactivity is also essential, where the exercise scenario must adapt to the trainees' responses in real time.

An "open" international exercise: Caribe Wave

The annual Caribe Wave exercise organized by UNESCO and coordinated by NOAA aims to advance tsunami preparedness efforts in the Caribbean and adjacent regions. Any participant from the 48 partner countries, whatever their level of responsibility (from the prefecture of the area to the elementary school) can register on the platform, tsunamizone.org, and thus have the detailed scenarios on D day. The scenarios are edited by the Pacific Tsunami Warning Center (PTWC). During the March 11, 2021 exercise, two scenarios were proposed. The first simulates a tsunami generated by a magnitude 8.0 earthquake located on the Enriquillo–Plantain Garden fault zone in Jamaica. The second is based on a tsunami generated by a magnitude 8.5 earthquake along the subduction zone north of the Lesser Antilles. One of the key points is to test the communications between the PTWC, the Tsunami Warning Focal Points (TWFP) and the National Tsunami Warning Centers (NTWC). The response to the scenario provided by the PTWC is at the discretion of each country. The involvement of each country varies but the lack of involvement of local authorities does not prevent local actors from engaging in actions of citizen awareness. The caribewave.info website provides exercise and evaluation guides.

3.2.4. *Communication and media awareness*

In addition to testing classic crisis management systems, we also conduct crisis communication exercises. This aspect is developing and has become a discipline in its own right. Future crisis managers need to train themselves to manage the upward and downward information, that is, the information that arrives in the crisis cell and also the information to be produced externally towards the population, the media, etc., and this passes through "media training", training on press releases and press briefings, etc. As an extension of these new considerations, it is also important to mention the use of Social Media in Emergency Management (SMEM) to obtain information on the current situation and transmit instructions to the population during an event. Like the VISOV association, for example, which supports decision-makers in crisis situations in this approach, this opens up innovative perspectives to further improve the realism of crisis exercises[15]. For its exercises, IMT Mines Alès is developing TwitterLike, a tool for simulating the social network Twitter in order to train crisis managers in SMEM: among other things, it relies on a high frequency of messages broadcast simultaneously containing different types of information that users must qualify and prioritize with respect to the situation encountered.

3.2.5. *Stress management*

Crisis management training is as much psychological as it is technical. Stress management, or rather the habituation of deciding and acting in a stressful situation, is part of the training of crisis managers and should be extended to the population. What could be better than an evacuation exercise in case of flooding in heavy rain! It is also accepted that the exercise must include a degree of uncertainty. The well-tested scenarios that are repetitions of procedures well known to the actors (called drills or acquisition of automatisms) cannot be considered learning exercises in crisis management since we have said that crisis by definition falls outside of the predefined frameworks of the ex-ante planning.

The stressful conditions are caused, as we saw in the previous chapter, by the uncertainty as to how the crisis will unfold and the presence of danger (terrorism, risk of explosion, collapse, etc.) and also by the harshness of climatic conditions, the lack of sleep, food or drink[16], isolation due to the lack of means of communication, etc. As of 2005, the Paris Fire Brigade trained to manage hyperterrorism in collaboration with other Parisian services (criminal brigade, anti-terrorist section, SAMU (Emergency Medical Service), first aid associations). The learning process concerned the management of uncertainty, the implementation of detection, identification, decontamination and rapid evacuation techniques, and inter-partnership conduct to find a common mode of action. The scenario was made credible by a complete lack of information for actors. The Prefect of Police, accompanied by his Director of Public Security, the Brigade Commander, his Chief of Staff and the Director of Facilitation knew where and when the exercises were taking place. This was accompanied by obvious stress, the sole purpose of which

---

15. https://inhesj.fr/publications/lirec/reseaux-sociaux-et-gestion-de-crise.

16. These conditions are often experienced by first responders after a natural disaster who must rely almost entirely on their own resources to survive and respond.

was to put the responders (firefighters, first aid workers, criminal brigade, anti-terrorist section) in conditions as close as possible to reality, that is, facing deliberate violence in an urban area transformed into a theater of war with the risk of encountering suicide bombers capable of blowing themselves up in the middle of a crowd. In the aftermath of the attacks of November 13, 2015, all government services had been training for 10 years in order to limit the impact of such a tragedy.

3.2.6. *A community-wide exercise*

A simulation exercise was conducted in June 2019 in the Paris region in Alfortville requiring the activation of its Communal Command Post (*Poste de Commandement Communal*, PCC). The objective for the municipality was to test the operationality of its Communal Protection Plan (*Plan Communal de Sauvegarde*, PCS) through the simulation of a slow kinetic flood caused by the Seine. The secondary objectives were to evaluate the coordination of the municipal team, interdepartmental relations, and the continuity of the community's activities by considering a massive evacuation of the population.

In order to meet all of the objectives set, the organizer[17] first solicited internal resources from the municipality in order to identify, upstream, certain vulnerabilities that would be interesting to integrate during the development of the exercise scenario. Indeed, the construction of a crisis exercise scenario is based on characteristics linked to the hazard (risks, previous events, etc.), to the territory (vulnerability of the stakes, etc.), to the different tools available to the municipality (internal and external tools used as decision aids, etc.), and also to the particularities of the crisis unit that will have to face this simulated situation (identification of certain profiles, possibly missing competences, experience, etc.).

Different actors were then made available on the day of the exercise to participate in its facilitation and to simulate the different interlocutors of the communal crisis cell. Present were the Flood Forecasting Service (*Service de Prévision des Crues*, SPC), the National Police, the Departmental Fire and Rescue Service (*Service Départemental d'Incendie et de Secours*, SDIS), the Interdepartmental Civil Defense and Protection Service (*Service Interministériel de Défense et de Protection Civiles*, SIDPC), ENEDIS, GRTgaz, VEOLIA, the EPTB Seine Grands Lacs, the Red Cross and journalists to simulate media pressure.

17. In this case, a specialized research firm called "Riscrises".

It is from its crisis room that the commune took part in this five-hour exercise, observed by two people equipped with observation grids. The evaluation is divided into two main entities: the observers, directly in the room, who observe the way the crisis unit organizes itself, distributes tasks, makes decisions, communicates, etc. while the facilitators focus on the relevance of the calls they receive. Visual animations in augmented reality were broadcasted directly on the screens of the PCC to increase immersion. Before the debriefing, a self-assessment questionnaire is distributed to each cell member to collect their overall feeling on the exercise, the points that bothered them, their feeling of immersion, their strong points, etc. These data are used for the cold debriefing, in addition to the other data (tools, hot debriefing).

The debriefing, which took place 45 minutes after the end of the exercise, enabled the trainer to identify good practices observed in the cell, some of which could be reviewed, while trying to understand why some of them had encountered difficulties. The trainer, with the help of the observers, must distinguish between difficulties that are related to the exercise itself and those that are related to the crisis management itself. One of the difficulties was to resolve the blockages related to the duration of the exercise. The municipality, generally confronted with the rather slow momentum of the phenomenon, was destabilized by the choice to condense the temporal scale in order to simulate, in five hours, what corresponds in reality to one month of event management. Some participants appreciated leaving their comfort zone; others did not. Concerning the crisis management itself, the exercise showed the importance of the presence of the Director of Emergency Operations (*Directeur des Opérations de Secours*, DOS) in the crisis cell. The trainer must avoid, at all costs, leadership of the debriefing being taken over by the person in charge of the cell, as the symbolism of hierarchical superiority risks inhibiting the voice of the other members. The identification of two distinct and diametrically opposed modes of operation of the crisis cell, explicitly one with the DOS and one without the DOS, led to a major recommendation: the person acting as authority over the rest of the group should be located in a separate room so as not to inhibit the activity of the cell. To avoid this type of problem, we can consider, for example, a "situation" cell with the people in charge of the different cells and a "decision" cell, for example the mayor's office. This case is a revealing example of the difficulties involving members with high levels of responsibility. It also shows that, through a crisis exercise, both operational and organizational recommendations can emerge.

3.3. Conclusion

A crisis upsets acquired knowledge. It calls into question acquired certainties and always surprises. Preparing for a crisis is a job for some, a duty for all. In this perspective, there are many forms of learning, from the analysis of past crises through feedback to exercises in "quasi real" conditions. Information technology, video and the media offer the means to compensate for the loss of knowledge that used to be provided by attachment to territories and the stability of populations. "Rational" knowledge of phenomena or of what to do is a first step, but it does not replace the experience of a crisis. Therefore, learning methods must be more and more realistic (virtual and/or augmented reality, indoor or outdoor simulation, etc.). Whoever the actor, from the layperson to the experienced professional, it is by practicing that the individual develops a strong capacity to anticipate the event in a group (team of rescuers, crisis unit, inhabitants in the field). However, it is necessary to go to the end of the process and accept that the exercise becomes increasingly unexpected and puts the player in a situation where only their own thinking skills will allow them to survive. The right reaction at the right time is not innate; it is learned in accordance with the adage: "Excellence is an art won by training and habituation" (Aristotle). Training is a guarantee of efficiency, it develops acuity, it teaches teamwork, and it teaches better anticipation. Exercises must be increasingly unannounced in order to test the procedures (and not the men and women who implement them) by accepting the surprises generated by the impromptu nature of certain reactions during exercises. It is the very essence of crisis to learn to be surprised. One of the keys is to bring together in the learning process, whatever its form, the experts of the territory, that is, the local community leaders and sometimes the inhabitants themselves, and the crisis management experts who are experienced in crisis exercises.

4

Crisis Management Resources

Jérémy ROBERT[1,3], Pauline GLUSKI[2], Sébastien HARDY[2], Pascale METZGER[2] and Alexis SIERRA[2]

[1] *Université Rennes 2, France*

[2] *Institut de Recherche pour le Développement (IRD), France*

[3] *Institut Français d'Etudes Andines, Lima, Peru*

Until the end of the 1990s, crisis management was considered a matter of civil protection: organizing the response to an emergency situation was a problem of logistics. In other words, it was a problem of action but not a research issue. It must be said that, as each crisis situation is unique, research was confronted with an epistemological obstacle (Morin 1972). The proliferation of feedback analyses has provided elements of understanding of crisis situations beyond their contingent nature (Godard et al. 2002; Gilbert 2005; Lagadec 2007a; Dedieu 2013). This research shows how crises reveal the vulnerabilities of societies: "The event resonates with its context. Like a cyclone that increases its power tenfold by drawing its energy from the seas it crosses, it feeds on all the problems and imbalances already present in the context in which it moves" (Lagadec 1991). It also helps us to guide the production of useful knowledge to improve response capacities in the face of future crises.

With the objective of contributing to this effort to understand crisis phenomena and their management, this chapter proposes a reflection on crisis management resources. To take the case of the Covid crisis in Peru, existing resources were adapted: makeshift tents were set up in the courtyards of hospitals. Private clinics were also mobilized. Alternatives were created: buildings constructed for the Pan American Games were transformed into hospitals, the Acho Arena was transformed

---

Also with the contribution of Robert d'ERCOLE, IRD

Territorial Crisis Management,
coordinated by Richard LAGANIER and Yyvette VEYRET.

into a shelter for the homeless. The Covid crisis has thus provoked extraordinary and specific challenges, requiring in turn extraordinary and specific responses and resources (Bergeron et al. 2021). Does this mean that it is impossible to anticipate these crises and the identification of the resources needed to address them?

The issue of crisis management resources is rarely addressed head-on in the literature. Three corpora of literature have been identified that provide specific and complementary insights. The first corpus is the feedback from experiences, which provides concrete elements on the impacts and disruptions generated by crises and on the resources mobilized to deal with them (Rey et al. 2019). In its analysis of the crisis caused by the 2010 earthquake in Haiti, Groupe URD (*Urgence Réhabilitation Développement*) highlights the impacts on the essential infrastructure (airport, decision-making centers, water and energy, etc.), the multiplicity of needs and the difficulties related to matching these needs to emergency aid actions (Grünewald et al. 2010). The work on the vulnerability of "critical infrastructures" extends these analyses from an anticipatory perspective. A second corpus of work deals with the planning of the allocation of means and resources in the event of a crisis. Based on modeling approaches (Özdamar and AlpErtem 2015; Paul and MacDonald 2016) or management approaches (Holguín-Veras et al. 2012), this work aims to improve the efficiency of logistics for emergency and humanitarian operations. From a crisis management preparedness perspective, Rodríguez-Espíndola et al. (2018) propose a model to identify the resources to be mobilized and the associated actors. Post-disaster needs are also the subject of original work, particularly through the analysis of social networks (Basu et al. 2017). A third corpus of work sheds additional light by focusing on local response capacities, from livelihoods (Gaillard 2013) and community resilience (Pasteur 2011).

Our reflection is at the crossroads of these approaches, and proposes to formalize a method for identifying crisis management resources in a territory, taking into account the multiplicity of scales and actors, as well as the diversity of potential dysfunctions. This reflection is based on several research experiences conducted by the team of the PACIVUR program (Andean program of research and training on vulnerability and risks in urban areas) of the IRD (French National Research Institute for Sustainable Development)[2]. The framework and method presented take up the challenge of trying to improve crisis management capacities while reducing

2. In particular, we will mobilize the SIRAD project (Lima, 2010–2011), "*Preparación ante desastre sísmico y/o tsunami y recuperación temprana en Lima y Callao*" (INDECI-PNUD-ECHO) and, to a lesser extent, the SIRV TAB project (Port-au-Prince, 2012–2013): "*Creation of a Resource and Vulnerability Information System for disaster preparedness, crisis management and preventive management of the territory of Port-au-Prince, Haiti*" (ECHO/HTI/BUD/2012/91016).

the vulnerability of territories, from a perspective of both fundamental and applied research. Section 4.1 presents the evolution of the issues within the PACIVUR program that led us to consider crisis management as a research object, and then addresses the conceptual implications. Section 4.2 presents the approach to the identification of crisis management resources, first distinguishing the different types of resources, before detailing the methodological aspects. Section 4.3 returns to the interest of an approach based on crisis management resources.

4.1. The reason for a shift from risk to crisis

4.1.1. *Why be interested in crises?*

Launched in April 2006 under the direction of Robert d'Ercole, the PACIVUR program is a follow-up to the research program "Information System and Risks in the Metropolitan District of Quito" developed in Ecuador between 1999 and 2004 within the framework of a collaboration between the IRD and the Municipality of the Metropolitan District of Quito (d'Ercole 2004). PACIVUR has a regional ambition, opening up to two other Andean countries, Peru and Bolivia, while maintaining a presence in Ecuador. The central objective of the program is the production of knowledge on urban vulnerabilities from an applied research perspective and based on collaborations with operational actors.

While the program conducted in Quito was mainly interested in risk management from a prevention and urban planning perspective, PACIVUR is initiating a research project on crisis management in an urban environment. By showing the limits of the Peruvian civil protection, the earthquake of Pisco (200 km south of Lima) of August 15, 2007 has, in fact, influenced this dynamic. Beyond this event, the interest in crisis management is justified by a set of observations drawn from the international bibliography and field experience: on the one hand, an intensifying urban vulnerability (poverty, social exclusion, uncontrolled urbanization), and, on the other hand, a generalized failure of preventive planning (exposed spaces that are largely urbanized, the limits of preventive systems, undesired effects of protection works), particularly in southern countries. At the same time, the importance of the *preparedness* phase in international forums is being observed (Revet 2011). However, if the reinforcement of crisis management capacities appears to be a possible alternative, civil protection mechanisms often appear to be inoperative (organizational and coordination problems, a lack of forecasting and preparation, non-existent or ineffective local management). Scientific production in this field does not seem to meet the needs of managers either, which are often poorly defined. Indeed, work on crisis management remains relatively limited, particularly in the field of geography (d'Ercole et al. 2009).

This discrepancy between scientific production and operational activities represents a challenge that also invites the development of new questions. Within the framework of PACIVUR, an exploratory research is conducted on the spatial and territorial dimensions of crises in three Andean capitals. In 2010, the reflections were applied in Lima with the realization of the project SIRAD "*Système d'information sur les ressources pour l'attention aux désastres*" (Information System on Resources for Disaster Attention).

This project was carried out by the PACIVUR team in collaboration with the NGO COOPI (Cooperazione Internazionale), financed by ECHO (European Commission Humanitarian Office) and supervised by the UNDP (United Nations Development Program) and INDECI (National Institute of Civil Protection of Peru) as part of a program of preparation of Lima and Callao against the occurrence of an earthquake of great magnitude and/or a tsunami. It aims to analyze the response capabilities and vulnerabilities of the Lima metropolis, which introduces a new problematic angle to the approach of risks.

Faced with this challenge, the main originality of the SIRAD project consists of creating a database on crisis management resources. The approach is based on two fundamental aspects: firstly, it is a question of focusing on crisis management resources, while initially ignoring the hazards. Secondly, it is about positioning itself on a metropolitan scale – on the whole urban agglomeration – in order to produce useful knowledge for public authorities. The project is based on the principle that the identification of these resources and the analysis of their vulnerability can provide useful knowledge to ensure the functioning of the city in the face of a disaster. It differs from previous studies (and those that are usually mobilized at the international level) that are focused on hazards (seismic zonings), on the estimation of damages and the number of potential victims (PREDES 2009) or on institutional and organizational aspects, for example, in terms of protocols. In this sense, it also requires an original conceptual framework.

4.1.2. *From stakes to resources: conceptual framework*

4.1.2.1. *Risk analysis by stakes*

The conceptual framework developed to analyze crisis management is largely based on work conducted in Quito as part of the program "Information System and Risk in the Metropolitan District of Quito". The initial and common objective of this research is to go beyond the limits of the classical risk approach (defined as the intersection of a hazard and a vulnerability) in order to improve its scope and operational potential.

In Quito, the research strategy was to focus the analysis not on hazards and hazard-prone areas alone, as is commonly done, but on the entire city and what it would be important to protect, that is, what is essential for the functioning of the city and could be affected. This is what has been called "major stakes" (or *lugares esenciales* in Spanish) (d'Ercole et al. 2002). The approach consists of first defining these major stakes, analyzing their vulnerability and then synthesizing this information to analyze the vulnerability of the territory (d'Ercole et al. 2004). This makes it possible to focus attention directly on what is of interest to policy makers, thus making the research more operational. Risk analysis starts from the territory, not the hazard. This approach inexorably leads to a number of conceptual changes (d'Ercole et al. 2005). The first of these is to place the stakes at the center of the definition of risk. Risk is then defined as "the possibility of losing what is valued". Stakes are defined as "what is valued", and "the possibility of losing" corresponds to vulnerability. Thus, there are stakes, of varying levels of vulnerability, that can be lost, and this is what constitutes a risk. The other conceptual consequence concerns hazards, which become one form of vulnerability among others. They are no longer opposed to vulnerability but included within it.

4.1.2.2. *From stakes to resources*

The shift from prevention (in Quito) to crisis (in Lima, then in Port-au-Prince) translates into a shift from the major stakes of the territory for urban functioning to the essential resources for the management of crisis situations (Table 4.1).

Decision and intervention	Emergency health
Water supply	Telecommunications
Energy supply (electrical and fuel)	Shelter areas
Transport and communication routes	Places for storing rubble
Food supply	Economic activities

Table 4.1. *Crisis management resources for the SIRAD project (d'Ercole et al. 2011)*

As soon as we are in the field of crisis, the term "resource" is preferred to that of "stake", in particular because it reflects the active character of the element considered. A resource is mobilizable and useful for managing a crisis. It concerns elements that will allow the situation to be managed and ensure the continuity of urban functioning. To speak of a resource also implies the existence of a demand to be met: the problems and specific needs caused by the disruptive event as well as by the dynamics of the crisis itself.

4.1.2.3. *Defining crisis management from a territorial perspective*

The spatial and territorial dimensions of crisis management are an object of research in geography that describes a particular crisis, which is neither a past crisis, nor a particular scenario. The problem of the spatial and territorial dimensions of the management of crisis situations proposes thinking about a crisis in general terms, that is, to consider the preparation for a crisis situation, whatever the physical, political or social phenomenon that is going to "cause a crisis". Projecting oneself into a major crisis requires thinking about dysfunctions at all levels of the territory, linked to multiple interrelations and dependencies, in a context of generalized uncertainty. The analysis of a territory's crisis management resources – located at different scales, from local to international – makes it possible to anticipate these dysfunctions, at least in part.

To grasp these dimensions, the proposed conceptual framework formalizes the relationship between the supply of resources and the demand in a territory to be rescued. The management of a crisis situation consists, in very concrete terms, of mobilizing resources to meet the needs in the spaces affected at the time of the disaster (the demand). In this logic, there are at least two different types of spaces created by crisis management: on the one hand, the spaces directly impacted by the disaster, which are to be rescued as a priority, and, on the other hand, the "resource spaces" for crisis management, that is, those that have the material and decision-making means for rescue and recovery (d'Ercole et al. 2009). Crisis management consists of linking these two types of spaces.

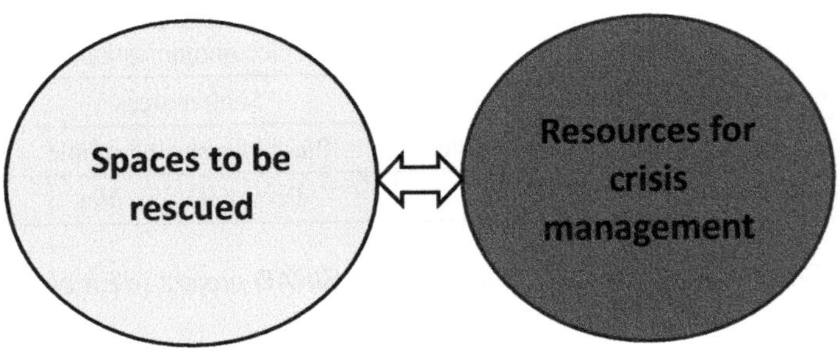

Figure 4.1. *Spatial dimensions of crisis management*

For the first type of space – the spaces to be rescued – the spaces that are a priori the most vulnerable, therefore, those that are most likely to be affected, are those that are considered. In particular, these are areas where a vulnerable population from a socio-economic point of view is concentrated, those that are highly exposed to hazards (highly seismic areas or areas that can be flooded by tsunamis, for example) or those that present accessibility difficulties. The identification of crisis management resources implies projecting oneself into an unprecedented situation that breaks with the usual functioning of the urban system.

This conceptual framework therefore makes it possible to build useful knowledge for crisis management. Indeed, even if we cannot determine a priori the extent and location of the damage, we can list, locate and characterize the crisis management resources, and we can determine a priori the most vulnerable places. Crisis preparedness, if these spaces are known, is able to reduce uncertainty regarding the management of the crisis, in particular, by acting on the location of crisis management resources. The identification, mapping and analysis of the spatial and functional relationships between these two types of spaces are therefore useful for emergency preparedness. The eminently geographical question of the location of crisis management resources in the territory, particularly in relation to the population distribution of the most vulnerable spaces and activities, makes the construction of a georeferenced database very useful. The analysis of the spatial and territorial dimensions of crisis management should make it possible to highlight situations that are a priori critical, that is, that suggest serious management difficulties in the event of a major crisis. This is the case, for example, of very vulnerable areas that do not have resources. This is also the case for disputed territories, which are subject to jurisdictional conflicts or illegal occupations, in which the presence of public authorities is deficient and difficult (Sierra 2016). Conversely, the analysis can identify crisis management territories that are a priori better off because they have local crisis management resources.

4.2. How to identify crisis management resources?

4.2.1. *Different types of crisis management resources*

In order to identify crisis management resources, the central hypothesis on which the reasoning is based is a very general and simple principle: in a crisis situation, the usual elements of the territorial system are degraded, to varying levels, and are only partially functional. This is a strategic hypothesis in the sense that it allows us to reflect on what crisis management resources can be with regard to the general characteristics of major urban crises: generalized dysfunction, overstretching capacities, multiplication of actors and a general imbalance between resources and needs (Godard et al. 2002; Robert 2012). The interest and coherence of this positioning is to consider the generalized dysfunction of the normal system as a hypothesis in itself, without needing to know the cause.

Crisis management resources are the elements that are useful, or even essential, to this management. Resources can be distinguished according to their type (tangible or intangible), their function (normal or specific crisis management operations) or their geographical scale (for a neighborhood or for the entire urban area). To identify and prioritize the crisis management resources in Lima, and to validate the information, research relied on thematic workshops bringing together specialists in each theme and representatives of institutions involved in crisis management.

4.2.1.1. *Human and material resources versus intangible resources*

Human and material resources, which are identifiable and locatable on the territory, can be concretely mobilized to manage crisis situations. The SIRAD project focuses on 10 types of resources (Table 4.1), which is possibly an incomplete list (see, for example, the case of urban soil, which is considered as a resource in (Robert et al. 2014)).

There are also intangible resources (which are therefore difficult to locate), such as professional skills, legal procedures, agreements and protocols, emergency operation plans and budgets. These resources enable the mobilization and optimization of the operation of physical resources. Management capacity or privileged relationships between different actors also fall into this category. Networks (professional, community or family) also play an important role in crisis management. Social organizations have been the focus of much attention from North American disaster sociologists, and *communities* are now considered to be the key element in risk and disaster management in the English language literature, particularly that which deals with the notion of resilience (Pelling 2003).

4.2.1.2. *Normal, alternative and crisis-specific resources*

Three types of essential crisis management resources can be distinguished: those that are essential in normal situations, alternative resources and resources that are specific to crisis situations:

– Normal functioning resources: these are resources that allow the daily functioning of the territory, outside of any disturbance, such as hospitals or a water purification plant. We can also mention the main road network, airports and ports, as well as refineries or the electricity network. All these elements are essential both for the functioning of the territory in normal conditions and for managing a crisis situation. In these cases, the identification of crisis management resources starts from the identification of the "major stakes" of normal urban functioning.

– Alternative resources: these correspond most often to a degraded functioning. They can only be occasional in time and space. To take an example, we can consider that there is no real alternative to the public network to supply water to the 10 million inhabitants of the Lima/Callao agglomeration. Nevertheless, in a temporary way, some wells and reservoirs are able to supply water and to temporarily and locally substitute for the public network, assuring a degraded but essential functioning. They are, therefore, alternative resources to the operation of the normal system. However, these elements depend on specific equipment that requires other resources, for example, energy for pumping wells, which implies access to electricity (or a generator and a stock of fuel). Here, we see that crisis management resources raise the question of the complexity and interdependence of systems, which are sources of vulnerability. A crisis management resource never functions alone but always depends on other networks and equipment that participate in the functioning of the territorial system.

– Another type of alternative to the water supply network is the fleet of water trucks and the filling stations for these trucks. In Lima, most of the water trucks are privately owned, which raises questions about their possible mobilization (thus a legal resource, i.e. immaterial) by public authorities in a crisis situation. Moreover, this alternative, for it to be operational, requires that the supply points of the tankers work, that there is fuel and that the roads are clear. The proper functioning of alternative crisis management resources is therefore subject to a series of dependencies, which are therefore likely to make them vulnerable. With regard to tankers, it can be noted that they are a normal mode of supply for certain neighborhoods or dwellings that are not connected to the public network, but are marginal on the scale of the agglomeration. We could therefore consider that neighborhoods that are not dependent on the public water system will not be affected. The situation may be quite different, however, since the tanker trucks will, in all likelihood, go to supply the rich neighborhoods that are deprived of water. In this case, it is not the earthquake itself that will suspend the water supply to these marginal neighborhoods but the rules of the market, that is, the law of the highest bidder.

Reflection on alternative crisis management resources to normal operation therefore leads to questioning the daily operation, particularly marginal, which presupposes, first of all, a good knowledge of the territory.

– Resources specific to crisis situations: these are resources that will enable new needs to be met. These resources do not play an essential role in normal situations, but become very important in times of crisis. They include, for example, crisis management and coordination institutions such as emergency operation committees, operational relief agencies such as fire departments, evacuation routes, field hospitals, NGOs, humanitarian aid storage centers, rubble disposal areas, heliports, etc. We can also consider all that is necessary to restore the resources of normal operation (companies, equipment, personnel, etc.) as specific and essential resources for the management of crisis situations.

The transition from the status of an element of normal operation to that of a crisis management resource is sometimes accompanied by a change in function. This is typically the case for space, when a recreational park is transformed into a shelter or when a vacant space is mobilized to deposit the rubble produced by earthquake. It is also the case when a cancer treatment center becomes a primary care center. In this case, the change of function is accompanied by a despecialization in order to adapt to the crisis situation. The mobilization of these specific crisis management resources depends on the situation and the needs caused by the crisis. It is especially significant when a crisis occurs, when the normal functioning capacities are exceeded and therefore when it is necessary to resort to exceptional measures.

To ensure water supply in a crisis situation, especially in a densely populated area and when the crisis is prolonged, decision-makers must develop several tactics in parallel for them to be efficient.

The main tactic is to restore the water supply system as soon as possible. Indeed, even if it is possible to distribute drinking water rations for a few hours to a few hundred, or even a few thousand disaster victims, this tactic is not tenable in the medium and long terms and with tens of thousands of disaster victims. It implies a heavy logistical burden: whether it is transporting bottles of drinking water by truck from emergency aid warehouses, or distributing drinking water directly to the victims by tanker truck.

When the crisis situation is prolonged, the water needs of the disaster victims increase, not only for drinking but also for hygiene: washing, laundry, often participating in the cleaning of the shelter spaces; sometimes cooking. The normal water supply system must be able to take over quickly. However, when it is damaged, decision-makers can mobilize alternative resources. If their role is temporary and often restricted to certain spaces in a normal situation, in a crisis situation, they can constitute an essential support to re-establish the functioning, even if it is degraded, of the water supply system.

In the Bolivian agglomeration of La Paz-El Alto, there are many springs: the water is not drinkable but it is of good quality, enough to be used as washing water or as water to be potabilized via filters placed in tanks. In fact, the water from the springs is already used on a daily basis for these different uses by poor populations and/or those not connected to the drinking water distribution network of the Pacenian agglomeration (Hardy 2013).

If water storage reservoirs are lost during a disaster, spring water can temporarily and occasionally overcome the shortcomings of the normal system, as was the case during the 2016–2017 water crisis (Le Gouill et al. 2019) (Figure 4.2). There is therefore an interest for decision-makers to identify these alternative resources upstream to understand how to mobilize them. For example, the location and flows of springs must be known. It is also necessary to know the availability of tankers, whether they will be able to easily access the springs to pump water, whether tanks and potabilization filters are available, etc. Managing support resources therefore implies preparation in advance of the crisis and knowledge of the territories, hence the interest in georeferencing this information in selective databases and sharing it between the various crisis management actors. It is also a way to integrate something that is considered informal into the normality of the supply system (Hardy 2020).

Figure 4.2. *The authorities of the municipality of La Paz have been using tanker trucks regularly since the water crisis of 2016–2017 to get water from springs for uses that do not require potabilized water (source: photograph by S. Hardy)*

Box 4.1. *Alternatives to normal operation that become key resources in a crisis situation: the case of water supply in La Paz*

4.2.1.3. *Geographic scale*

Defining crisis resources implies asking the question of the geographical scale of crisis management. The elements that can constitute resources are different depending on whether we are looking at the metropolitan scale or the local scale. A health center can be a critical crisis resource at the neighborhood level, while being of no particular importance at the metropolitan level. A pharmacy kit can save an individual's life; a major hospital, on the other hand, is essential to the population.

Furthermore, the scale of management of a resource does not necessarily correspond to the political–administrative level where it is located, and this resource is not necessarily managed (or even mobilizable) by the decision-maker in charge of the jurisdiction in question (Metzger et al. 2013). Thus, the mayor of a district does not have the capacity to directly mobilize humanitarian aid resources, even when they are located in their jurisdiction. In other words, depending on the type of resource, the most appropriate management scale varies. Some resources can be managed at the metropolitan level (e.g. major roads, telecommunications, hospitals). Others can be managed – at least in part – at the local level (alternative water supply to the public network, shelters, rubble dumps, first aid) in order to optimize the response to disaster populations and the restoration of infrastructures, in time and space.

Each scale has its own crisis management resources. A hierarchy of resources exists within a given field. In the food sector, a distinction can be made between national wholesale markets, local retail markets, secondary markets and semi-wholesalers, whose catchment areas correspond to distinct orders of magnitude and territories. During a crisis, each of these markets is likely to have a different role depending on the territorial scope of the disaster. In the same way, in parallel with metropolitan crisis management resources, city dwellers mobilize local resources that are immediately accessible: dispensaries and first aid centers, neighborhood wells and reservoirs, gas stations and mechanical workshops, municipal police, local radio stations, etc. These resources are both complementary to the resources of the metropolis and are also used by the local population. They are both complementary to the resources of higher levels and of greater proximity. They depend on local decisions, but these are more or less autonomous. The management of these local resources is generally the responsibility of the municipalities, but the central state remains present through its decentralized services and administrations, as do the large public service companies. The same applies to large NGOs or churches and their charitable organizations.

The organization of decision-making in the event of a crisis also takes place at different levels, through crisis cells (with local cells as replicas of the national cell). In Peru, there is a National Council for Disaster Risk Management, a regional council, a provincial council and a municipal council, each level with its own emergency operation committee. At the local level, around the mayor, the heads of the administrations and deconcentrated services of the state, as well as the local economic and social leaders, meet. However, this pattern is not the same everywhere, and in more centralized states, the mayor's role is devolved to the local representative of the central government (the equivalent of the French prefect), who can, incidentally, do without such a broad-based action committee. Situations of transition to more decentralized systems, promoted by international organizations, are moments of great tension between local authorities. Claude Gilbert described this precisely in the case of the Nîmes floods of 1988, when the decentralization laws had recently been enacted (Gilbert 1992). The management of the Covid-19 pandemic in Tunisia offers a very good example. In the Greater Tunis agglomeration, some municipalities took emergency measures in accordance with the recently enacted 2018 law on local authorities. These decisions were taken up a posteriori by the central government, which has a strong presence in the capital and has imposed a control a priori of local decisions at the same time as the health state of emergency. It is thus slowing down a process of decentralization, maintaining its emergency response resources (civil protection, police, emergency medical services, etc.) under the authority of the governor. Crises can thus reveal or exacerbate conflicts between levels of government, and, more generally, between the different actors involved in resource management.

Pachacutec is a peripheral territory located in the extreme north of the Lima-Callao agglomeration, depending on the municipality of Ventanilla. It was formed in 2000 as a result of the large-scale arrival of thousands of Limenians from the south and center of the metropolis, in search of a place to live. The army, after having opened the roads, made the earthworks and transported the population, let them build their houses and organize themselves with NGOs and some state services. The population has also developed its own emergency resources. Used on a daily basis, we hypothesize that they are alternative means in the event of a crisis and a halt in public services, the latter being little present. The population travels locally by mototaxi, a motorized tricycle, which, according to the drivers, can transport injured people and small pieces of equipment. The neighborhoods have loudspeakers, whose use is primarily commercial, but which, during fires, have been used to mobilize residents and local solidarity. Finally, some food services are developed locally from national programs for the poor, and residents have developed their own mutual aid networks, particularly through associations of people from the same regions.

Box 4.2. *Emergency resources at the neighborhood level in Pachacutec (Lima, Peru)*

The capacity to mobilize local crisis management resources also arises when it is the responsibility of private or at least non-state actors, such as businesses, NGOs, churches and their charitable organizations, or simply city dwellers themselves. Consequently, the change in scale can result in a substantial change, both in terms of quantity and in the nature of the resources. If they have overall the same function, they do not necessarily have the same legal status, the same social acceptability or the same mode of use. The case of Pachacutec illustrates, in an extreme situation, what the local level can offer as a resource for mobility, communication and services. These resources, with a strong component of informality, are viewed with suspicion by the authorities, who do not control them. Mototaxis (auto-rickshaws), whose drivers have a bad reputation, often considered to be former delinquents, are not recognized in their urbanity in normal times, and are generally confined to certain peripheral or marginal parts of the Limenian territory (Plas and Sierra 2018). However, the drivers themselves consider that they could be useful in the case of an emergency and we can only recall what Parisian cabs contributed as an emergency service at the very beginning of World War I.

4.2.2. *Methodological aspects*

How can crisis management resources be identified, and above all, go beyond a simple list to produce useful knowledge about the territory's response capacities and vulnerabilities? Inventories of crisis resources may exist in a formal and localized way, for example, in emergency operation plans or in official files of municipalities or ministries. However, this only concerns a limited number of them, and most institutions do not have a localized inventory of their own resources[3]. How can this information be completed? How can it be organized and prioritized? Below, we present some methodological aspects concerning the identification of crisis management resources. The specificity of the approach lies in the materiality of the objects considered to be resources and not in the phenomenon (hazard, event) that interrupts normal functioning. In this sense, the approach is similar to the analysis of critical infrastructures (Box 4.3), with which it shares the same philosophy, while insisting on the importance of a precise definition of the objects and the scales of analysis.

First thought of in the United States at the end of the 1990s from a national security perspective, following several terrorist attacks (in New York in 1993, in Oklahoma City in 1995 and in Tokyo in 1995), critical infrastructures were first listed in 1996 by the Presidential Commission on Critical Infrastructure Protection (PCCIP 1996). This commission specified that these infrastructures are so vital that "their incapacity or destruction would significantly weaken the defense or economic security of the United States"

3. The French Communal Protection Plans (*Plans de sauvegarde Communaux*) mention "human and material resources" which correspond to crisis management resources. The development guide recommends making an inventory of these resources (a list, telephone numbers based on the Orsec plan model), in particular, for all public resources, but also for private or associative resources. However, the guide does not propose locating them, nor analyzing their vulnerability (thus, in the event of a disaster, it is not impossible to see the fire station or the operational crisis management center completely out of service – see the flood in Nîmes in 1988 and the recent flood in the Roya).

(Galland 2010). About 10 infrastructures are cited, including telecommunications, energy and water supply systems, transportation, health, etc. The objective is to identify these infrastructures on a national scale, analyze their vulnerability and develop protection strategies. These activities require collaboration between the government and the private sector, which owns or operates most of these infrastructures. The attacks of September 11, 2001 reinforced these previously limited initiatives. The concept was taken up again in Europe in the 2000s (notably following the Madrid attacks in 2004 and London attacks in 2005). A Green Paper "on a European Programme for Critical Infrastructure Protection" was written by the Commission of the European Communities in 2005 (CCC 2005).

Critical infrastructures are generally defined as infrastructures whose loss can lead to major social, economic and political dysfunction and disorder. Reflections on the vulnerability of these infrastructures, particularly concerning large networks, are now becoming central to the management of large-scale crises and risks (Michel-Kerjean 2003). However, despite a growing interest, there is a certain vagueness concerning both the definitions and methods for identifying critical infrastructures (protect what?) and threats (protect against what?) (Galland 2010). Some authors conclude that there will never be a definitive definition (La Porte 2006). Some focus on transnational infrastructures (Bouchon 2009), others on national ones. Similarly, while a multi-hazard approach is sometimes adopted, approaches are clearly focused on terrorist threats, particularly since the events of September 11, 2001 (Ocqueteau 2007).

A geographical approach can provide elements of solutions to the vagueness of this definition. It seems fundamental to integrate the territorial scale and the social point of view, which will determine what is critical infrastructure and for whom, for example, identifying resources at the scale of an urban agglomeration and for the public authorities in charge of managing a crisis.

Another bias concerns the definition of what is critical according to the threat. For example, if we consider the terrorist threat, using the example given by Michel-Kerjean (2003), the entire water system infrastructure can be targeted by the introduction of a pathogen. It is obviously not a question of posting a guard at all the places that are likely to be attacked: this is economically impossible. If the entire network is taken into account, it is no longer possible to distinguish the key elements or to carry out effective protection measures. The first step is to ignore the threat and identify the key elements according to their role in the service's operation. In this way, we can try to limit the propagation of damage by interrupting the transmission of vulnerabilities and by thinking in terms of alternatives.

Box 4.3. *The logic of critical infrastructure*

4.2.2.1. *Starting from the territory and projecting into the crisis*

One of the objectives of crisis management is to restore the normal functioning of the territory. However, the crisis is at the origin of generalized dysfunctions that are difficult to anticipate. The analysis of the normal functioning of the territory (its major stakes and its institutional organization) is an essential but not sufficient step to identify the resources for crisis management. It is also necessary to project oneself into a crisis situation. It is necessary to imagine, for a given territory, how the resources of normal operation will be mobilized with regard to the potential effects of the crisis[4], as well as to think about the non-functioning or degraded functioning of these resources. The crisis will indeed provoke a reorganization of the resources of normal functioning, soliciting the mobilization of alternatives and relying on specific resources. This approach is consistent with the typology of crisis management organizations proposed by Russell Dynes (1970) and taken up by Arjen Boin et al. (2010), which shows that crisis management involves both specialized actors of daily emergencies, actors of normal functioning and exceptional actors. Feedback and *post-crisis* analyses also provide useful elements for this exercise, and show the diversity of the resources mobilized.

4.2.2.2. *Constructing a georeferenced database*

The construction of a georeferenced database aims to gather information on available resources, their location and capacities, and allows the prioritization of these resources. It represents an essential tool for decision-making in a crisis situation (Box 4.4). In fact, information on resources is, in itself, a resource (a resource that is neither identified nor located is difficult to mobilize and is therefore not really a resource).

The construction of a georeferenced database also allows the realization of certain analyses through a geographic information system (GIS). For example, we can cross-reference different types of information such as seismic zoning or accessibility. We can also analyze resources according to proximity criteria, or the adequacy between resources and vulnerable spaces in a territory. The database constructed within the framework of the SIRV TAB project carried out by the PACIVUR team in Port-au-Prince allows us to visualize the potential cross-referencing of information (Figure 4.3). Moreover, such a database can be useful for other purposes (this is the case of the database built within the framework of the SIRAD project, which was then mobilized for territorial planning purposes by the metropolitan authorities).

4. We can also recall, even if it seems obvious, that resources are not identified according to their level of exposure to a hazard.

The difficulties associated with the lack of information are evident in every major crisis. During the Ancash earthquake in 1970 (Peru), there were no maps of the affected areas to manage the crisis situation. A "basic intelligence" service was created to compensate for this lack within the new civil protection organization. In New Orleans in 2005, the circulation of information posed a major problem: "despite the prodigious means of information and communication, the Katrina fiasco showed the system's inability to provide the information needed for decision support in a timely manner and at the right places" (Lagadec 2007b). In Haiti, under the impetus of United Nations organizations such as OCHA (Office for the Coordination of Humanitarian Affairs), forums have been created to centralize the geographic information produced by the multitude of international agencies and organizations following earthquakes. It is possible to download base maps, political–administrative boundaries and databases of healthcare facilities (in "shape" format with metadata). Although the amount of information produced is impressive, this information is often only used to produce location maps, without any real analysis, hence, the need for upstream reflection in order to orientate the information to be constructed.

Box 4.4. *The value of georeferenced databases on crisis management resources*

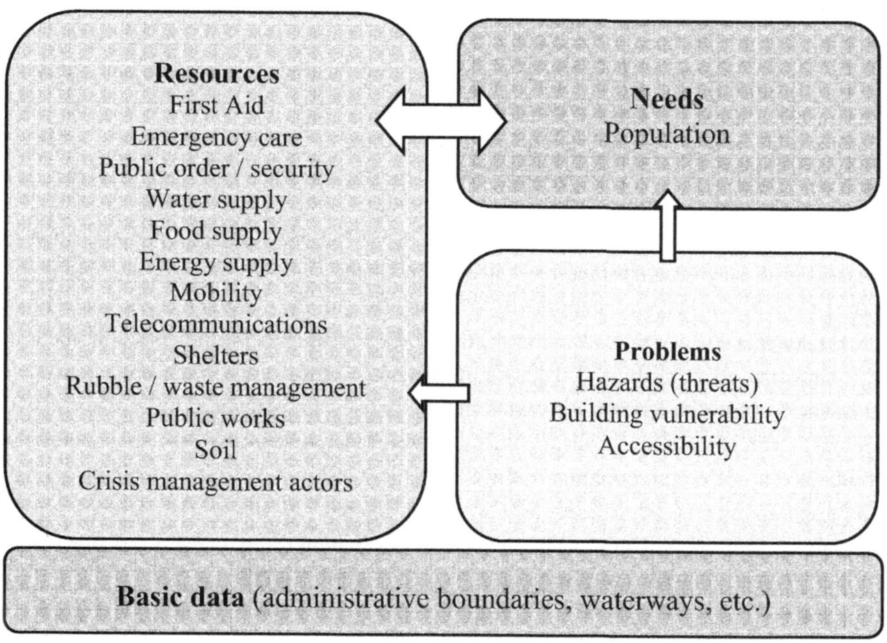

Figure 4.3. *Structure of the SIRV-TAB project database*

A georeferenced database is composed of information layers that contain a series of objects of the same nature, represented in the form of points, lines or polygons, and described with the same attributes (characteristics of the object). For example, in the SIRAD project database, the "healthcare facilities" information layer contains 529 facilities (objects), represented as points. Each healthcare facility is described with the same attributes (name of the facility, institution, number of beds, etc.). Metadata accompanies each information layer, including a general description of the content, the date of creation, the sources, the geographical scope, as well as the precise description of each attribute and the methodologies used.

Constructing a georeferenced database that can be used with a GIS requires a priori reflection on the nature, description and scale of representation of the objects. Each information layer must gather objects of the same nature. It is impossible to put two different objects in the same information layer, for example, ambulances and healthcare facilities, even if the hospitals have ambulances. In the SIRAD project, an information layer was created for each crisis management resource. In the case of ambulances and hospitals, two separate information layers were constructed, and the "care facilities" layer has an attribute that describes the number of ambulances in the facility. The objects were defined according to the identified crisis management resources. This is one possibility among others. For example, blood banks were isolated in a specific information layer, whereas it would have been possible to consider them in the attributes of the "care facilities" layer (all blood banks are located in a care facility). By creating an information layer, this resource is given more importance by making it more visible and is characterized as such, with its own attributes (e.g. type, available stock, etc.). Users of the database can then identify it more quickly.

The construction of the database is therefore not trivial and requires a real reflection on the status of these objects. It imposes choices, formalizing what is a resource and taking into account the relevant scales of analysis. By requiring rigor in the definition of the status of the objects manipulated and the structure of the data to be constructed, the tool also contributes to the formulation of problems and methods of analysis. Clarification of concepts, definitions and research objects also helps us to avoid inconsistencies. For example, J.P. Galland (2010) mentions, with regard to critical infrastructure, "inconsistencies due to inconsistent definitions: all of the transportation in the San Francisco Bay Area was covered by a single item [an information layer], whereas New York had entered 739 subway stations separately in the database. More importantly, many items (e.g. gas stations) were entered and filled in, or not filled in, depending on the information collector" (Galland 2010).

Box 4.5. *The construction of objects within a georeferenced database*

The construction of each information layer requires the collection of existing information, which can rarely be integrated directly into a database. It is necessary to homogenize them, to locate them and to complete the attribute information with other sources or with the help of surveys and interviews according to the analysis needs.

Beyond the difficulty of imagining what the resources that will allow a crisis situation to be managed might be, there are multiple obstacles to building a database on crisis management resources. For all the services and equipment managed by the private sector, knowledge of the location and characteristics of the equipment and infrastructure may be considered to be confidential, notably because of commercial issues. This is particularly the case for telecommunications in Peru, which is emblematic of a privatization process at the end of which the public authorities no longer have any resources of their own, apart from the emergency radio bases and the public radio channel. However, access to data is also complicated in the public sector. The services directly managed by public authorities are fragmented into a multitude of territorial levels and organizations, which represents a first difficulty in knowing the resources of each one. In addition, the armed forces and the national police have significant personnel and logistical resources, both for maintaining public order and for immediate recovery or humanitarian aid activities. This data is often very difficult, if not impossible, to obtain. The lack of information on the resources available to manage a crisis situation is, in itself, indicative of a territory's vulnerability.

4.2.2.3. *Prioritizing resources*

The elements that are actually useful for crisis management are, a priori, multiple and scattered over the territory. However, if all these resources are important, some are more important than others. A hierarchy of resources according to their level of importance can therefore be defined at a given scale. In the SIRAD project, three levels of importance – or "essentiality" – have been defined:

– Major crisis management resources: these are resources that will play a decisive role in crisis management for the entire territory, for which there are no or few alternatives, and whose loss is likely to compromise the response capacities for a significant part or even the entire territory and population.

– Supporting resources: these are resources that provide significant support and/or possible alternatives without replacing major resources. They do not have the same importance as major resources, particularly in terms of territorial coverage. The loss of a supporting resource results in a more local impact. More numerous and of lesser capacity, together they constitute an important resource for crisis management, making it possible to avoid saturation of the major resources. They are also resources whose mobilization presents foreseeable difficulties, in terms of costs or protocols.

– Other resources: a third level gathers other resources, which, although useful, are of lesser importance in crisis management or are very local.

This prioritization operation is all the more necessary since identifying all the resources in a territory is impossible, or even counterproductive, because the really important resources are drowned in the mass of information.

A first step in prioritizing resources is based on the definition of quantitative, qualitative and spatial criteria (see Table 4.2).

Types of criteria	Examples and description
Quantitative criteria	A resource is major because of its size: in terms of the number of beds in a hospital; the hierarchy of a water supply system (primary or secondary); the size of a space to set up a camp; the stock of a market.
Qualitative criteria	A resource is major when it has a particularly useful and/or uncommon quality: the presence of emergency, surgical or burn services in a hospital; the equipment of a well and the quality of the water available; the connections to the water and electricity network in a park planned as a refuge.
Spatial criteria	A resource is major because of its location in relation to other resources of the same type. This means taking into account proximity criteria, with regard to the spatial distribution of certain resources in the territory. A health facility may not be very important in terms of quantity or quality, but it may be the only one to serve a large area or be easily isolated in the event of a crisis. It can then be considered a major resource. It may also be a well in an area that is not served by the water network, etc.

Table 4.2. *Criteria for selection of major resources*

These different criteria are defined empirically. Quantitative criteria are the most commonly used, in order to make an initial hierarchy according to thresholds set by the researcher. Qualitative criteria allow us to attribute an importance to an element, according to a specific quality, which makes it an uncommon element with a high functional value. Finally, spatial (or location) criteria allow for consideration of the distribution of resources over the territory.

4.2.2.4. *Soliciting operational actors and experts*

The identification of resources and their prioritization can be greatly enriched by soliciting operational actors and experts, who will bring in their technical, political and social knowledge regarding the functioning of the resource. This approach makes it possible to compare the analyses carried out by researchers with the knowledge of operational actors. Through their function, experience and practice, they will provide new insights into the functioning of their sector (and level) of action. On the other hand, these specialists often lack a systematic and spatialized reflection on crisis management resources in a given territory. They generally focus on institutional, coordination or logistical aspects, without concretely identifying or locating the resources that can be mobilized[5]. The challenge is therefore to integrate their knowledge into the analysis.

4.2.2.5. *Application to emergency health resources in case of a major earthquake in Lima and Callao*

Different types of actors were solicited in workshops for the identification of emergency health resources: representatives of key institutions (such as the Ministry of Health) who have an overall vision, and who allow access to information; directors of hospital emergency departments, who have a concrete vision of the emergency and daily problems; doctors who intervened in Haiti or Chile following the 2010 earthquakes; firefighters, members of the National Institute of Civil Protection (INDECI), representatives of the health services of the municipality of Lima or associative and NGO actors. Whenever possible, we asked for a representative of the different territorial levels (local, municipal, provincial, regional, metropolitan and national).

Eight types of emergency health resources, with the primary purpose of caring for the injured, were identified (Table 4.3). The nature of the data to be constructed on these resources, that is, how to describe them in order to estimate their capacities, was also discussed. For example, it is imperative to know the number of beds in healthcare facilities, as well as the number of operating rooms, or the capacity of a blood bank or a drug warehouse. Using the experience and knowledge of some of the actors made it possible to avoid missing out on relevant resources.

This exchange process has been enriching in several ways beyond the identification of resources. First of all, it allowed for a contribution in terms of knowledge and experience sharing. These workshops not only allow the researcher's viewpoint to be compared with that of the operational staff, but also allow for

5. The use of cartographic documents is uncommon, leading to analyses that are too general and ubiquitous, and do not reflect the specificities of the territory.

exchanges between different actors who do not often have the opportunity to meet. They also have practical advantages in terms of disseminating research, making contacts and accessing information.

Healthcare facilities	Blood banks
Health sector actors	Country hospitals
First aid groups	Potential hospital expansion areas
Drug warehouses	Ambulances

Table 4.3. *Types of emergency health resources analyzed in the SIRAD project in Lima (Robert 2012)*

Table 4.4 presents the eight types of resources identified by classifying them according to their role in a major crisis situation. This hierarchy, although indicative, is fundamental. Indeed, if we consider all the resources in the database without distinction, it is difficult to propose an analysis of the care system. The mapping of the number of beds per health facility and the selection of major resources is more telling (Map 4.1). It shows a reduced number of essential resources for crisis management. It also shows a strategic central area, contrasting with poorly equipped peripheries.

This diagnosis already provides elements for reflection in terms of vulnerability: an unequal spatial distribution coupled with quantitative and qualitative limitations, such as a relatively small number of beds on the scale of the conurbation, saturation of services on a daily basis, or the concentration of specialties that are useful for emergency healthcare in a very limited number of establishments. These difficulties are exacerbated by problems of coordination between actors.

4.2.2.6. *Analyzing resource vulnerability*

Once major resources are identified, a further step is to analyze their vulnerability. These analyses can focus on major resources, allowing for more specific and detailed analyses. Several forms of vulnerability can be analyzed, from exposure to hazards, susceptibility to damage (e.g. vulnerability of buildings), preparedness, dependencies on other resources or the existence of alternatives (Robert and Metzger 2004). This work was, for example, carried out within the framework of the SIRAD project in Lima for the 23 major hospitals of the capital. Geophysical measurements, coupled with an analysis of response capabilities and level of preparedness, highlighted the most vulnerable major hospitals, which should receive special attention (Robert 2012).

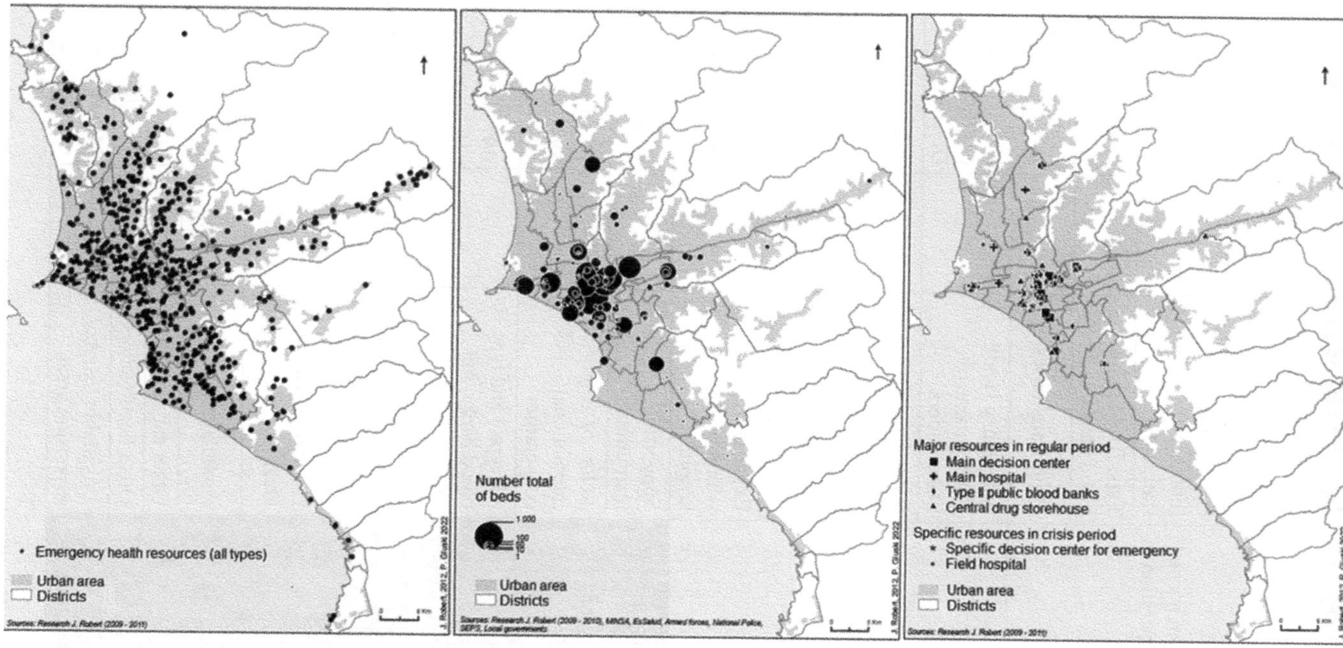

Map 4.1. *All emergency health resources in Lima Callao; number of beds by health facilities; major and specific emergency health resources in crisis situations*

			Quantity of objects	Indicative description
Major resources	From normal operation	Major hospitals	23	10,391 beds (68.3%)
		Main actors of the decision	9	Directions of Minsa and EsSalud
		Fire department (number of fire stations)	49	4,271 firefighters
		Type II and public blood banks	23	1,734 blood bags in stock (81.4%)
		Main warehouses	11	Central warehouses of Minsa and EsSalud
	Specific to the crisis	Emergency Health Management Stakeholders	7	NDMO, NDO and CEPECED, Central Fire Department
		Specialized first aid groups	9	365 rescuers
		Emergency warehouses	1	Specific warehouse of Diresa Callao
		Country hospitals	3	104 beds
		Expansion areas within major hospitals	180	Open spaces that can accommodate several tents
Supporting resources		Supporting institutions	153	4,805 beds (31.6%)
		Actors of international cooperation	14	PAHO, Red Cross, NGOs, etc.
		All type I and type II blood banks, but private	55	397 bags in stock (18.6%)
		Containers of medication for emergency use	15	15 containers for 100 people
		External expansion areas	14	Nearby public spaces (or private spaces provided for this purpose)
		Ambulances	110 (locations)	181 ambulances
Other resources		Other care facilities	353	16 beds (0.01%)
		Other health actors	19	Training centers
		Unsuitable external expansion areas	37	More distant and/or private spaces

Table 4.4. *Prioritization of emergency health resources (Robert 2012)*

4.3. The benefits of a resource-based approach

4.3.1. *Understanding the vulnerability of the territory*

Detecting the problems posed by the location of crisis management resources in relation to vulnerable spaces is a first way of addressing the spatial dimensions of crisis management. Two types of vulnerabilities can be identified by considering the spatial dimensions of crisis management:

– Vulnerability due to the spatial distribution of resources: where are the resources for crisis management located, how are they distributed over the territory in relation to the areas that are most likely to be affected? What proximity or distance, what ease of communication between vulnerable spaces and resource spaces?

– The degree of dependence of vulnerable areas on resources. In other words, what is the availability of crisis management resources within the most vulnerable spaces, taking into account, in particular, potential accessibility problems.

At the same time, analysis makes it possible to identify two types of spaces, knowledge of which is particularly useful for preparing for crisis situations:

– On the one hand, critical areas of crisis management, that is, areas that are likely to be in great difficulty, not only because of their own vulnerability (socio-economic, institutional, accessibility, etc.) but also because of their difficulties in accessing crisis management resources, which can be identified a priori;

– On the other hand, territories whose crisis management would be facilitated, because they are relatively autonomous and have the human and material resources needed to enable their own crisis management capacity.

This type of analysis makes it possible to highlight territories that are more or less vulnerable according to their crisis management capacities.

Moreover, the prospect of a major crisis requires us to think about urban dysfunctions and their consequences. These dysfunctions have multiple origins and illustrate the interrelations not only between different types of resources, but also between resources and the territory. The functioning of a particular resource necessarily depends on several other resources. For example, the proper functioning of hospitals depends on other urban services such as water supply, energy supply or telecommunication systems. However, the transmission of vulnerability from urban services to hospitals can be limited when the latter have alternatives and/or sufficient autonomy. In addition, crisis management resources meet certain needs and their malfunction may affect all or part of the urban territory. Damage to a major hospital at the city level can result in the loss of access to care for a large number of people. In turn, the vulnerability of hospitals depends on the number of potential patients that could cause them to be overwhelmed.

A territorial vision of crisis management therefore consists of taking into consideration not only the adequacy between supply and demand, but their interrelations, or even feedbacks. Preparing for crisis management must therefore integrate, on the one hand, the interconnection and interdependence of the various networks, infrastructures and equipment, and, on the other hand, the relationships between supply and demand in the territory. Thinking in terms of resources makes it possible to identify the phenomena of transmission of vulnerabilities and to develop strategies to interrupt the chain effects. Crisis management therefore implies thinking about the functioning of the territory as a whole.

4.3.2. *An operational interest*

The analysis of crisis management resources (particularly the work of prioritization) makes it possible to draw the attention of decision-makers and managers to the major resources (and, particularly, to the most vulnerable among them) in order to consider public policies for prevention and preparation for disasters, and to contribute to guaranteeing the functioning of these resources in emergency situations or their rapid recovery after the occurrence of the destructive phenomenon.

Identifying major resources has a triple benefit:

– For decision support at the time of a crisis, because it gives the authorities reference points in a context of general confusion and allows them to establish priorities. To manage a crisis, it is necessary, on the one hand, that decision-makers have a global vision of the needs on the territory, and, on the other hand, that they have knowledge of the location and characterization of the available resources. The relevance and effectiveness of the action depend on the scale of decisions (political and administrative), the knowledge of decision-makers and the location of resources. Knowing the resources available for each level of crisis management and what can be mobilized at another level, either by pooling (between districts, for example) or by hierarchy (from the national to the local level), can be particularly useful. This can help coordinate and prioritize interventions. Emergency plans could usefully include a localized inventory of resources that can be used to manage a crisis, specifying the actor(s) authorized to decide on their use.

– For the reduction of vulnerability: the identification of major resources allows the optimization of all actions and costs. For example, this allows vulnerability analyses to be concentrated on a limited number of elements, interventions to be carried out on the most vulnerable ones, or preparedness exercises to be carried out by mobilizing resources that are likely to play a key role during a crisis. Moreover, this approach offers a coherent framework that is capable of integrating the contributions of different disciplines. For example, the contribution of environmental sciences is essential for the analysis of the vulnerability of resources, more particularly the vulnerability of buildings, as well as for identifying the areas where the population is exposed to the different hazards (in addition to more social analyses). However, it is necessary to identify upstream the resources where attention should be focused.

– To understand the functioning of the territory and the dynamics of crises: the analysis of the resources of a territory allows us to formalize and produce knowledge on its functioning. By identifying the vulnerability of the interacting elements (resources, populations), the proposed approach makes it possible to highlight the flaws in the functioning and structural vulnerabilities of the territory, which are more apparent in times of crisis than in normal times.

4.4. Conclusion: thinking about a "risk–crisis–development" continuum

The resource-based approach to crisis management contributes to the production of useful knowledge for crisis management preparation, by emphasizing the importance of territorial dimensions. It also highlights the structural vulnerabilities of the territories, and opens up other avenues for action. This approach makes it possible to highlight the links between these structural vulnerabilities and the potential problems at the time of crisis, and how these links materialize in the territories. Crisis is therefore not seen as a more or less independent phase of risk management, but as part of a continuum of "risk prevention–crisis management–territorial development". Thinking about this continuum implies reformulating the problems, clarifying the links between the different phases and thinking about both the processes of building vulnerabilities over time and the transmission of vulnerabilities within the territory.

In fact, risk prevention measures necessarily improve crisis preparedness: reinforcing a hospital or better planning the supply of care in the territory reduces the vulnerabilities that are likely to fuel the crisis. Conversely, preparedness actions improve the functioning of the territory and are likely to reduce structural vulnerabilities: reinforcing coordination between actors and the autonomy of basic services in healthcare establishments, or involving social networks and citizen participation. In a way, it is a matter of socializing and politicizing crisis management preparation.

Acting on resources and their location also becomes a question of urban and regional planning, and of sectoral policies, which are not the responsibility of prevention policies, nor of preparation for crisis situations. The consideration of risks in land management is most often limited to avoiding the planning of facilities or infrastructures in hazard zones, and to limiting or even prohibiting new constructions. On the contrary, here it is a question of acting for prevention and preparation through an equitable management of the territory, that is, on a location of urban services and equipment that guarantees access, both in normal times and in crisis situations. In other words, the reduction of the territory's vulnerability can constitute a planning option that is capable of building a territory that is both fairer in normal situations and less vulnerable in the event of a disaster.

Thinking about a scientific object built on a "risk–crisis–development" continuum implies going beyond the temporal divisions imposed by action – the before, during and after. It is also a matter of incorporating the problem of risks and crises into that of the functioning of the territory and of development in general, as well as into that of inequalities (Metzger et al. 2014). Indeed, the processes of risk and crisis construction are inseparable from the modalities of territory production, and thus of development.

5

Post-disaster Recovery: Challenges and Resources

Annabelle MOATTY
UMR RECOVER, INRAE - Aix-Marseille Université, France

Disasters and the immediate post-crisis periods that follow are highly publicized; reconstruction does not receive as much attention. The recovery process, however, mobilizes considerable funds and energy over long periods of time, ranging from three to more than 20 years (Oliver-Smith 1991; Wisner et al. 2012; Su et al. 2020). Decisions made in the post-disaster period can have consequences on the development of societies and territories, as well as on the process of production or risk reduction several decades after the disaster (Vinet 2003; Gaillard 2010; Moatty et al. 2017; Davidsson 2020). Post-disaster recovery is a relatively recent concept in the francophone scientific literature on risk studies in geography. Internationally, research on recovering from disasters emerged in the 1970s with the development of multidisciplinary teams. The vast majority of works analyze the different aspects of a specific sector of the process (e.g. housing, land use, trauma and psychological consequences) without developing a systemic approach (Smith et al. 2007). Others have considered the recovery issue in a broader sense by treating it in parallel with issues of sustainable development (Berke et al. 1993; Berke 1995). These works mark an evolution of the concept of recovery, which is then understood as an opportunity to rethink the development of a society in order to make it less vulnerable, more equitable and put a more sustainable distribution and use of resources into place.

Territorial Crisis Management,
coordinated by Richard LAGANIER and Yyvette VEYRET.

The post-disaster period can indeed be a tipping point in the resilience process involving different forms of adaptations that span reactive adjustment and proactive prevention strategies to improve the response to future threats (Moatty et al. 2018). Resilience is a complex and dynamic phenomenon that refers to the triple capacity of systems: resistance during the crisis, the ability to recover and rebound immediately after the shock, and the ability to adapt and transform over time (Quenault 2013; Reghezza-Zitt and Rufat 2015). The development of the concepts of adaptation and Build Back Better (BBB) is contemporary with the advent of resilience (Moatty et al. 2018). This BBB ideal has become an international injunction, promoted by the United Nations in 2015 within the Sendai Framework for Action on Disaster Risk Reduction (Nations Office for Disaster Risk Reduction 2015). The damage and disorganization in the aftermath of a disaster act as an indicator and amplifier of the territory's vulnerabilities (Birkland 1998; Plein 2019). This highlighting of vulnerabilities is conducive to the development of mitigation actions, improving the resilience capacity of society, in other words, to "take advantage of the recovery process". This change in development pathways requires planning work that takes time. However, one of the challenges of the post-disaster period is also to respond rapidly to the needs of the affected populations. Faced with this difficult reconciliation of the short and long term needs, risk governance[1] is frequently restructured (Moatty 2020).

However, the preventive role of reconstruction can be questioned because of the persistence of risks, the difficulties of disaster victims to recover and the lasting deterioration of living conditions after a disaster (Nigg 1995; Olshansky et al. 2012; Lamond et al. 2015; Kaniasty 2020; Su et al. 2020). Disaster does not erase the past, and reconstruction does not start with a blank page free of legacy on which everything can be rebuilt better. On the contrary, post-disaster recovery builds on existing political, administrative, economic and socio-cultural structures (Berke et al. 1993; Alexander 2010; Aldrich 2012; Cretney 2017). Furthermore, disasters can be interpreted as an extension of everyday emergencies (Baird et al. 1975) and deep vulnerabilities (Oliver-Smith 1991; Wisner et al. 2004; Gaillard 2010; Magnan et al. 2016). Therefore, many obstacles (political, regulatory, technical, organizational, social) linked in part to the inertia of systems limit the capacity of post-disaster to be a preventive tipping point in the vulnerability pathways of socio-territorial systems (Whittle et al. 2012; Monteil et al. 2020). The interactions between development and

1. Risk governance refers to the set of actions, processes, traditions and institutions by which authority is exercised and decisions are made and implemented regarding the identification, assessment, management and communication of risk (IRGC 2005; Renn et al. 2018).

disaster are multiple and can be levers or pitfalls in the adaptation of development pathways of the affected regions (Anderson 1989; Lizarralde et al. 2009; Lyons 2009; Jha 2010; Moatty et al. 2017). In this chapter, which focuses on the challenges and resources of post-disaster recovery, we begin with a presentation of the different actors and temporalities of the process, modes of governance and regulatory tools associated with this period. We then analyze the combination of resources and constraints in terms of financing the reconstruction and recovery efforts. This is followed by a section on land use issues, in which we discuss the different territorial modalities of reconstruction. Finally, the identification of local variations in reconstruction allows us to conclude on the social and territorial inequalities and injustices generated by the decisions and actions taken during the post-disaster period.

5.1. The challenge of coordinating a multitude of actors with inadequate regulatory frameworks

5.1.1. *The actors and temporalities of the reconstruction process*

Post-disaster recovery involves all the actors in the socio-territorial system. It involves thinking on the scale of the disaster area and its "solidarity area", understood as the spatial extension of solidarity and mutual aid movements between disaster-stricken and neighboring areas (Moatty 2015). At the beginning of the reconstruction process, the central objective is to satisfy immediate needs for rehousing and initiate the resumption of public services and economic activities (see Figure 5.1). The key decisions then relate to the programming of credits and the financing of the increase in public spending, changes in land use rules, the setting up of a one-stop shop for businesses and individuals to facilitate their procedures, and coordination between the various aid mechanisms set up to ensure consistency and inclusiveness. The post-disaster period is composed of five overlapping periods. From emergency to recovery – through rehabilitation, restoration and reconstruction – managers and decision-makers mobilize and adapt financial, administrative and regulatory tools to drive and support recovery. Territories and social groups are not equal when it comes to reconstruction and experience variations in the duration of the different phases (color gradients in Figure 5.1). The reconstruction phase has its own temporality, but it is not disconnected from the development of the territory of which it is a component. Thus, while most actions and decisions are concentrated in the first few months and years, the consequences of reconstruction choices can last for some 20 years.

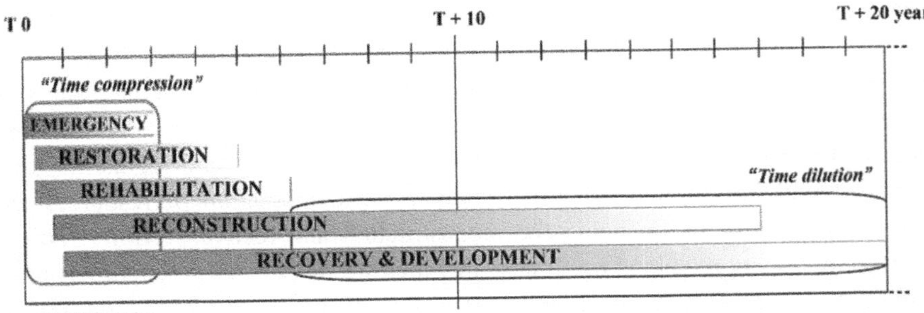

EMERGENCY
Emergency safety and relocation, victim registration, removal and clean-up, damage and needs assessment

RESTORATION
Waste sorting, restoring accessibility, temporary relocation solutions, implementing business continuity, restructuring governance modes to schedule funding

REHABILITATION
Processing of aid applications and funding programming, rehabilitation works, demolitions, feasibility and impact studies for reconstruction projects

RECONSTRUCTION
Feedback on crisis management, permanent rehousing solutions, implementation of works, accompanying measures for economic recovery

RECOVERY & DEVELOPMENT
Follow-up of the works, implementation of the "big" reconstruction sites, resumption of daily activities

Figure 5.1. *Phases of the post-crisis period and main associated actions (produced by the author)*

The early stages of the process are marked by the amount of decisions to be made and actions to be taken quickly to avoid letting temporary precarious situations drag on. The actors are faced with a "time compression" (Olshansky et al. 2012; Moatty et al. 2017). This is in contrast to the "time dilution" that characterizes the end of the reconstruction process. During this second period, reconstruction actions are carried out in parallel with daily missions and classic governance modalities are implemented again. In between, there is this slightly suspended time when the credits have been programmed and the work is underway, but the society and its territory have not yet returned to "normal", unimpaired functioning. It is in this phase that the first crisis experience feedback is initiated, during which the objectives and methods of the different actors are confronted. Decision-making is often contentious because conflicts exist, latent tensions may resurface, and interests may diverge or even be opposed between stakeholders (Gaillard 2010; Moatty 2015; Moatty et al. 2019). "Disasters can cross jurisdictional boundaries, create the need to undertake unfamiliar tasks, change the structure of the response organization, [and] trigger the mobilization of participants who do not ordinarily respond to crisis situations" (der Heide 1989).

	Actors	Crisis	Emergency	Restoration	Rehabilitation	Reconstruction
	United Nations	–	F	F	F	F
	European Union	–	F	F	F	F
	State and government services	F/D/A/	F/D/A/I	F/D/A/I	F/D/A//I	F/D/A/I
	Regions and departments	F/E	F/E	F/D/A	F//D/A	F/D/A
Public	Mayor and intercommunality	D/A/I	D/A/I	F/D/A/I	F/D/A/I	F/D/A/I
	Public establishments	-	-	F/D/A	F/D/A	F/D/A
	Local authorities	-	-	F/D/A	F/D/A	F/D/A
	Army	A	A	-	-	-
	Civil security	A	A	-	-	-
	Reinsurance	-	-	-	F	F
	Insurance	-	-	F	F	F
	Business and Industry	A	A	F/D/A	F/D/A	F/D/A
Private	Vital network operators and critical infrastructures	A	A	F/D/A	F/D/A	F/D/A
	Media	I	I	I	I	I
	Associations	A	A	A/F	A/F	A/F
	Individuals and communities	A	A	A	A	A

Table 5.1. *Public and private actors and their main missions during the different phases of the post-disaster recovery process (inspired by the French operation)*

Insofar as the crisis results from the fact that the response capacities of societies are exceeded (Lagadec 1991; Lagadec et al. 2008), the post-disaster period implies an evolution of daily missions to meet the recovery challenges. Some actors are present and active throughout the whole process; others are only present during certain phases and/or specific missions (see Table 5.1). Table 5.1 lists the main reconstruction actors and their missions (F: funding, D: decision, A: action, I: information) during the different periods of the process.

Depending on the constitution and administrative structure of states, the actors involved, their missions and their decision-making or strategic influence vary. The main difference in terms of management is the decentralized or centralized nature of power and competencies (in urban planning, the environment, social and economic policies, etc.), which gives more or less influence to local actors. A constant in the process is the interdepartmental restructuration, which is systematically carried out in the form of interdepartmental groups, committees and programs.

These decompartmentalizations of the governance modes can take many forms, such as the Pinatubo reconstruction commission in the Philippines (Gaillard 2010), the Napier Committee in New Zealand (Hill et al. 2013), *the Rekompak* program in Indonesia following the 2004 earthquake and tsunami, and the 2006 and 2010 eruptions (Moatty et al. 2017), or the appointment of a minister of reconstruction in Japan after the disaster of 2011 (Scoccimarro 2013). In France, the *Cellule ou Mission Interministérielle de Reconstruction* (CIR-MIR, Interdepartmental Reconstruction Unit or Mission) fulfils this role, even though it does not directly integrate the disaster victims in the decision process (Moatty et al. 2018). Under the authority of a local representative of the state (the Prefect in France), all the ministries are brought together to study the requests for aid (to individuals) and subsidies (to companies) and program the allocation of funds. This management, which has advantages, particularly in terms of saving time, fluidity and coherence in the decision-making and programming of descriptions, nevertheless hardly involves citizens and remains very *top-down*. In addition, the evaluation of the links between emergency, rehabilitation and development highlights several major weaknesses in intervention mechanisms at the global level (Brusset et al. 2011). First and foremost, is the centralizing logic that underlies this restructuring of governance modes, which can be vulnerable to corruption and loss of funds (Rodolpho 1995; Gaillard 2008).

Managing the transition phase from crisis management to temporary and then permanent strategies is a critical threshold within which the risk of falling into poverty is significant. It is during this transitional period that solidarity mechanisms play a major role, from the micro-local to the national level, which make it possible to temporarily and punctually compensate for the shortcomings of public action linked in part to the general state of disorganization (Moatty et al. 2020). NGOs can act more flexibly than states, even though their interventions are subject to the national law of the country in which they deploy their aid programs. They intervene at the request of states, and also benefit from a right to interfere (which is debated within the profession) that allows them to intervene without the state's agreement in the event of a human rights violation (Herlemont-Zoritchak 2009). NGOs and associations occasionally fill in the gaps (on specific territories, on a specific area such as education, access to water, health, etc.) and the inabilities of public policies to respond to the needs of populations. On the contrary, recent disasters such as the 2004 tsunami in Indonesia (Vinet et al. 2011), the 2010 earthquake in Haiti (Grünewald 2010; Jabouin et al. 2012) or the 2015 earthquake in Nepal (Cook et al. 2018), to name a few, have demonstrated that the actions of NGOs can also delay long term recovery by keeping a territory on life support (Grünewald et al. 2010). It also happens that, due to a lack of knowledge regarding local specificities and intermediaries on site, reconstruction projects are not adapted and struggle to achieve their goal of supporting the population. This is particularly true in contexts of significant disparities within the population, where the lack of consultation with the resource actors of the territory and communities leads to inequitable planning of aid because it ignores local conditions, culture and capacities (Comfort et al. 2010; Moatty 2015).

5.1.2. *Legislative and regulatory frameworks for recovery*

Post-disaster recovery suffers from the lack of a single framework – or at least integrated plans – that allows for the coordination of the process, as there is for crisis management, for example. Each actor, therefore, operates with its own strategic documents – such as Business Continuity Plans, for example (Moatty et al. 2020) – or adapted procedures. The laws and regulations that govern post-disaster situations are those that govern daily life and land-use planning, and are proving inadequate to deal with "time compression" on the one hand, and maintaining planning efforts over the long term, on the other hand. At the international level, to meet the assistance needs of developing countries, the World Bank, the European Union and the United Nations developed the *Post-Disaster Needs Assessment* (PDNA) in 2006. This framework provides technical and financial assistance to support the adaptation of reconstruction policies that integrate the principles of sustainable development. However, the first versions of the document required a level of technical expertise and skills that not all developing country governments necessarily have, especially in post-disaster situations, it was therefore coupled with training and local capacity building mechanisms (Jeggle et al. 2018).

In France, despite the inadequacy of procedures to act quickly and resist "time dilution", the law does not hinder reconstruction and recovery, but it does put decision-makers at risk. Obtaining the authorizations and exemptions can be a lengthy process, which discourages stakeholders when faced with slow and cumbersome procedures. However, they are a guarantee of the appropriate use of public funds and are frequently adjusted post-disaster. For example, regarding public contracts in the local authority code, an exemption is systematically granted for an average period of three months. This makes it possible to shorten the response time for calls for tender and simplify the procedures for selecting companies. Insofar as these adjustments are implemented by the interdepartmental structures in charge of managing recovery (see *Supra*, the CIR/MIR), they can be qualified as voluntary transgressions of the law with the endorsement of the state. Other forms of adaptation exist, such as the "misuse" of existing procedures that are not contrary to the law (Moatty et al. 2018). The modifications are made at the margins, taking advantage of legal uncertainties, and are the result of the inventiveness of stakeholders (such as the extension of the eligibility criteria for aid to reduce substandard housing in order to help the most precarious to finance repairs to their housing). A final type of adaptation concerns changes in the law to meet the needs generated by the post-crisis situation, such as the decree to remove the ceiling on state aid to municipalities (from 80% to 100% to enable the most modest communities to recover).

The law does not prevent reconstruction; on the contrary, it authorizes and supervises modifications that go towards reducing vulnerability. Experience has shown that when there is no legal injunction, reconstruction is carried out in the same way so as not to increase the existing vulnerability. In terms of adaptation, the question arises in a different way depending on whether they are dealing with engineering or protective structures, or housing. For engineering structures or dikes, the need to rebuild as quickly as possible to maintain their functions (sometimes vital for the territory) is not conducive to changing the characteristics of the structures, especially when they are only partially damaged. Sizing new structures, using new materials, etc. requires preliminary studies that take at least several months. In addition, there are constraints on certain structures, such as their classification as historical monuments or UNESCO World Heritage sites, which sometimes make it difficult to reconcile heritage conservation and risk prevention.

Faced with the lack of knowledge of the induced effects, positive or negative, the absence of certainty about the gains of a different reconstruction, the lack of criteria to judge an improvement and the decision to rebuild differently raises the question of the responsibility of decision-makers. Decision-making in a post-crisis situation comes up against multiple uncertainties. For institutional stakeholders, the uncertainty relates, in particular, to the consequences of decisions made in an "out-of-framework situation" (regulatory and legislative) on issues that evolve and interact. For the institutions, it is a question of deciding quickly and without having all the cards in hand. Identical reconstruction is faster, less costly and offers a sharing of responsibilities. Indeed, by following in the footsteps of what was done in the past, today's decision-makers share a "historical responsibility" with yesterday's decision-makers (Moatty 2015).

Therefore, improving the quality of reconstruction is done at the expense of its speed, and the need for speed in the early stages is a prerequisite for effective recovery, since it is not a question of allowing crisis and precarious situations that are supposed to remain temporary to drag on. Several factors explain these delays, such as the availability of resources (human, financial, technical and material), as well as the capacity of the actors to mobilize them effectively.

5.2. Financial resources for reconstruction and the weight of solidarity in individual recovery

International donors contribute to the financial effort of reconstruction by supplementing government resources. The restructuring of post-disaster governance is therefore also based on the need to mobilize available resources and coordinate external aid in order to recover.

5.2.1. *Government resources*

At the government level, financial resources for reconstruction are drawn from budgetary reserves and development finance. These funds are mobilized in the post-crisis emergency to meet the immediate needs of the affected populations and must therefore be released and disbursed very quickly (Mahul et al. 2010; Moatty et al. 2017). Reconstruction funding can be classified into two categories: those that can only be mobilized after the disaster and which are most often aimed at responding to the emergency and do not require detailed planning to access the funding, and those that can be mobilized preventively and are conditional on the development of a specific action program. These include budget reallocation, domestic and external credit, tax increases and private donor support. Prevention resources that can be mobilized for reconstruction include disaster funds, budgetary reserves, debt financing mechanisms and risk transfer mechanisms (mainly through insurance and reinsurance).

At best, state reserves are in the range of 2–5% of government spending (Mahul et al. 2010). However, they are neither earmarked for responses to disasters nor are they unlimited. Recourse to these reserves is compromised in the event of multiple simultaneous crises (natural, technological, economic, sanitary, etc.) or in the event of the accumulation of these crises over time. Moreover, systematic recourse to budget reallocations jeopardizes development programs that have often required years of preparation (Moatty et al. 2017). However, it is from these funds that assistance for disaster victims is taken, particularly for the most vulnerable and/or marginalized populations who do not have access to other sources of economic aid for recovery (insurance, technical or partial unemployment, etc.).

In France, the state is solvent and has the resources to help communities respond financially to disasters. The various mechanisms allow for the compensation of disaster victims and the reconstruction of damaged property, through the private insurance system, or under the supervision of the state, and the compensation of community property not covered by the insurance systems. Local elected officials have access to funds, the *Fonds d'Action Extérieur des Collectivités Territoriales* (External Action Fund for Local Authorities), created in 2013, which provides emergency assistance. The missions taken in charge range from the removal of wrecks to embellishment projects, repair, cleaning, or even support for populations and medical care. In addition, the *Fonds de Solidarité Logement* (Solidarity Housing Fund) can be activated to help tenants in difficulty to pay their rent. It is complemented by a General Interest Program (*Programme d'Intérêt Général*) to fight against substandard housing and energy insecurity, or to adapt housing for people with reduced mobility (Moatty 2015). The repair and reconstruction of uninsured and uninsurable public property can be subsidized by different mechanisms depending on the amount of damage. If the amount of damage exceeds six million euros, program 122 provides for the request of the public disaster fund, and if the amount is lower, it is the Solidarity Fund for Local Authorities which is requested.

The French system, built on the principle of national solidarity[2], is based on the mutualization of risks and damage costs through insurance and the Cat' Nat' procedure (Barraqué et al. 2020). The Cat' Nat' procedure is an insurance system

2. The principle of national solidarity reflects the involvement of the state in social and economic life and dates back to the economic crisis of 1930 and the period following World War II. It is closely associated with the concept of the welfare state and has led to the creation of several health and social protection institutions based on a system of compulsory insurance organized by the state (e.g. social security and unemployment insurance).

supervised by the state, designed to compensate victims of losses and damages caused by disasters not covered by traditional insurance contracts, provided that a natural disaster decree has been published in the official journal. This system is based on the solidarity of policyholders who all contribute 12% of their home insurance policies and 6% of their motor vehicle policies.

5.2.2. *International aid directed mainly to developing countries*

At the international level, several emergency and reconstruction aid funds exist (see Table 5.2). The main funders are the World Bank and the European Union. These mechanisms are complemented by the Global Bank Disaster Risk Reduction and Reconstruction Fund. The International Bank for Reconstruction and Development (IBRD) provides loans, guarantees, and analytical and advisory services to middle-income and creditworthy poor countries in Latin America, Asia, Africa and Eastern Europe. Through the Global Fund for International Development and Reconstruction, the IBRD ensures that low-income countries (GDP per capita of less than $1,135) have access to capital on favorable terms (very long term, interest-free loans), in larger volumes, with longer maturities, and in a more sustainable manner than the market. The IBRD primarily finances itself by selling bonds on the international capital markets. The objective is to help the poorest countries meet the cost of getting out of poverty, including the cost of providing essential services in education, health, water and sanitation.

At the regional level, we can cite the *Caribbean Catastrophe Risk Insurance Facility*, which brings together 16 Caribbean countries, and the *Pacific Disaster Reserve Fund*, which was created on the same model a decade later. With the support of the World Bank, these structures offer services comparable to insurance to deal with major disasters such as earthquakes and tropical cyclones. Other international actors have also developed emergency and reconstruction financing, such as the Center for Global Philanthropy in disaster response, which heads the Global Recovery Fund. In addition, private actors have a growing role in financing reconstruction on an international scale. By increasing the number of insured persons, the aim is to reduce the burden on governments at the same time. In the long term, measures to support the recovery of populations can be complemented by the use of solidarity microfinance, particularly to diversify sources of income (Lecomte 2008).

Funder/institution	Fund name and acronym	Date of creation
World Bank	International Bank for Reconstruction and Development – IBRD	1944, post-World War II
	Global Facility for DRR – GFDRR	2006
European Union	European Bank for Reconstruction and Development – EBRD	1991, post-Cold War
	European Union Solidarity Fund – EUSF	2002 – following flooding in Eastern Europe
Global Bank Group	Global Bank Disaster Risk Reduction and Reconstruction Fund – DRRRF	2014
	Global Fund for International Development and Reconstruction – GFIDR	2015
Member countries and donors	Caribbean Catastrophe Risk Insurance Facility – CCRIF	2007
Member countries and donors	Pacific Catastrophic Risk Assessment and Financing Initiative – PCRAFI	2016
Center for Global Philanthropy	Global Recovery Fund	2019

Table 5.2. *Non-exhaustive list of international sources of reconstruction funding*

5.2.3. *The influence of solidarity mechanisms on the recovery of individuals and communities*

The actions of institutions, insurance companies and associations providing aid to victims are complemented by solidarity reactions that are organized on an ad hoc basis to help disaster victims. The actors of post-disaster solidarity present a variety of statuses: NGOs, charities, local elected officials, public services, foundations and corporate foundations, individuals, amateur clubs (especially sports clubs) and service clubs. These actors take charge of the whole process: initiating calls for donations; preparing the action; collecting financial and material donations; transporting and distributing aid. Solidarity is expressed through different types of actions (financial and material donations, administrative assistance and psychological support, provision of human resources) and is implemented at different levels. Given the deteriorated conditions of accessibility, it is first a solidarity of neighborhood, of relations of geographical proximity, which is set up. It is a "physical local solidarity", since it is characterized by actions to help with repairs and cleaning, partly constrained by accessibility difficulties (Moatty et al. 2020). Then, with the arrival of outside help, a "solidarity of support" completes the first one (by sharing experiences and listening) by remobilizing the network of relations of individuals beyond the neighborhood (Moatty et al. 2020).

Beyond the specific case of the post-disaster period, solidarity can be broken down into three categories: (i) family solidarity, which is established in a system of kinship or contractual institution such as marriage; (ii) civil solidarity understood in the sense of a voluntary investment guided by ethical values associated with the feeling of being morally concerned by the situation; and (iii) social solidarity understood in the sense of social law, which therefore evokes a union of individuals "in the name of specific common interests" (Supiot 2015). This typology implies that the different categories of solidarity are variable in time and space: contracts can be broken, compassion and empathy can fade, and interests can change.

Although they are not the only actors of solidarity, the associations are the most representative, whether they existed before the disaster or were created on the model of the disaster victims' associations[3]. Foundations are also active, private organizations that mainly operate with private funds, thus guaranteeing their independence from the state. In France, the *Fondation de France* is the main player in this category, financing aid to the affected and victims through the restoration of living conditions (housing rehabilitation, support for economic recovery, and support for educational and cultural activities for young people, particularly through the financing of local associations). Service clubs (such as the Lions or Rotary Club) are also active in post-disaster solidarity. Organized on a network model that is often international, they help fund charitable or educational projects and are both recipients of aid and donors.

The last decade has been marked by the increasing mobilization of social networks used as a means of collecting donations from the general public. No legislation applies to these pools, which facilitates the collection of funds. These sums are either directly transferred to the disaster victims according to several eligibility criteria, or are used to buy construction materials, basic necessities or specific products such as food and baby equipment. In-kind and hand-to-mouth donations are favored by many organizations during the emergency and vulnerable population assistance phases. However, there are several challenges. First, it is often difficult to sort the donations received, both according to quality criteria and to the needs of the disaster victims. Donors and volunteers, who are not always on the ground and do not always know the territory, have good intentions, but they sometimes miss their goal (such as, for example, the sending of winter clothes to tropical territories). In addition, hygiene

3. Disaster associations aim to defend common interests, provide psychological support and set up leisure activities for the general public or specific audiences such as children (Moatty 2015).

problems linked to the time lag between the collection and distribution dates can affect a solidarity action. In the case of the 2004 Asian tsunami, the most obvious example is the donation of medicines: the World Health Organization declared that 25% of the shipment was out of date (Philippe et al. 2007).

5.3. Land resources and territorial restructuring

Considering the land resources for post-disaster reconstruction means integrating both the disaster area to be rebuilt and its "solidarity area". Different territorial strategies are put in place and can be broken down into two main approaches: reconstruction on site or relocation. The latter is a radical way of recomposing the territory, and redevelopment through changes in the way buildings are occupied also influences settlement dynamics. Numerous constraints and prerequisites are imposed on the choice of reconstruction territories, from the problems of land fragmentation to legal blockages, including the absence of a land register and the lack of knowledge of land ownership by managers and decision-makers. In addition, the disaster itself can be structuring from a morphological point of view, since spaces disappear due to erosion phenomena on the coastline and riverbanks, for example. At the individual level, this translates into situations of households who had legally occupied land before the disaster, but find themselves without land and sometimes unable to prove the exact location and legality of their destroyed property (Moatty 2015).

5.3.1. *Relocation of areas deemed too dangerous for population settlement*

Relocating single assets or entire villages is a radical way to reduce the exposure to hazards. However, it poses legal and financial questions for expropriating and rehousing the affected households. In France, Article 61 of the law of July 30, 2003[4] allows the mayor to acquire property amicably when the lives of the occupants are threatened, and the *Fonds de Prévention des Risques Naturels Majeurs* (FPRNM, also called "*Fonds Barnier*", Prevention Fund for Major Natural Risks) can be called upon to buy the property amicably or to expropriate its occupants if no agreement is reached. This situation is extremely rare and the rule is to buy the property out-of-court (Moatty 2015). The pre-emption procedure completes these arrangements. This is a right that allows local authorities with urban planning responsibilities to purchase land at a price set in consultation with the Prefect. When the property is bought, the rule is that the site is protected and there is little

4. See: https://www.legifrance.gouv.fr/loda/id/JORFTEXT000000604335/2020-12-13/.

possibility of revaluing these areas, except in the case of flood or submersion risks, where the land can be reclassified as a flood expansion zone.

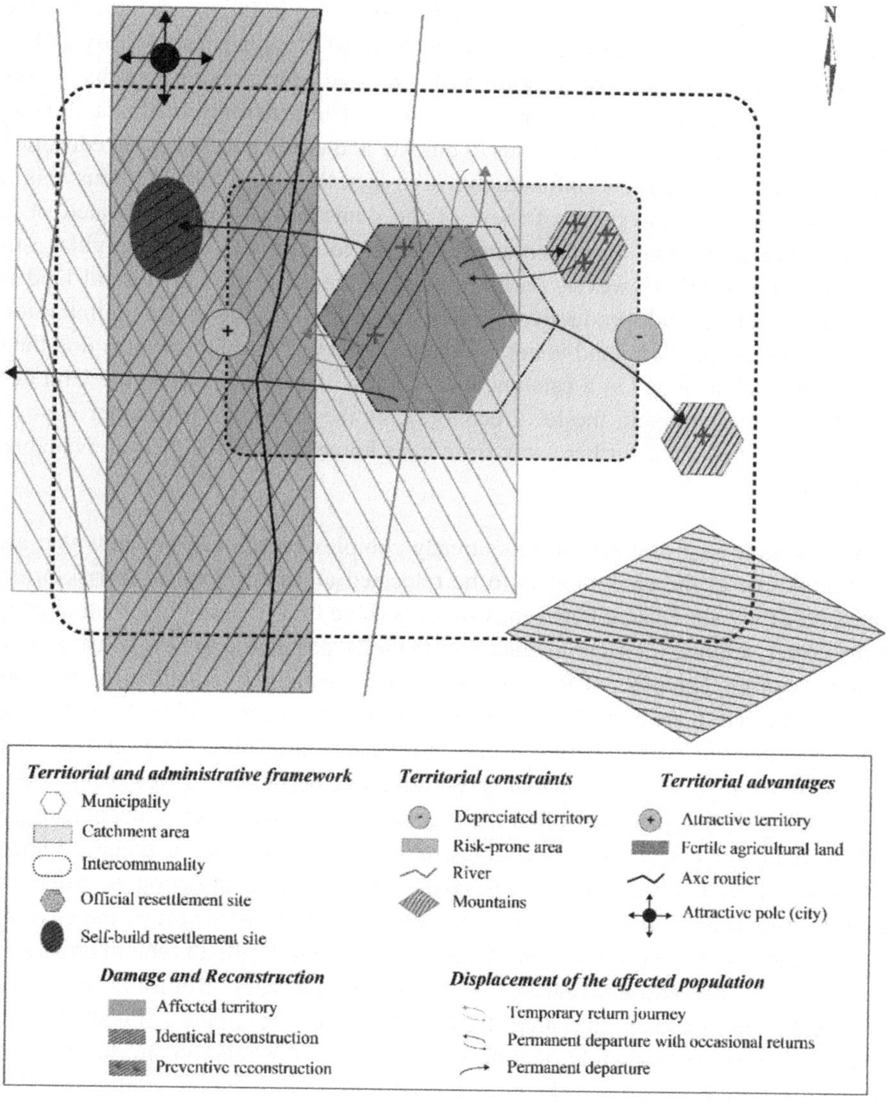

Figure 5.2. *Schematic representation of territorial constraints in the reconstruction process (adapted from Moatty (2015)). For a color version of this figure, see www.iste.co.uk/laganier/territorial.zip*

Relocation also raises the question of land availability outside of risk-prone areas. Inspired by chorematic, Figure 5.2 presents a synthesis of the socio-territorial constraints that restrict the possibilities of recomposing the territory. The case represented in Figure 5.2 is based on a study carried out in Indonesia following the post-eruptive *lahars*[5] of the Merapi volcano in 2010 (Java Island) (Moatty 2015). The affected area (in dark gray, Figure 5.2) is located near a major road that serves the attractive pole of the region: the nearest city. The proximity of this axis is an advantage since it allows communities to quickly and easily access health, education and commercial services and infrastructures. In fact, the land adjacent to this axis is highly prized. From the point of view of the communities to be relocated, it is therefore preferable that the resettlement site be located west of their village of origin (the hexagon). However, on both sides of the road, the agricultural land is particularly fertile and productive (pink rectangle). For the host communities, however, selling this farmland is not a sustainable or viable solution since it is their main source of income. On a catchment area scale, the sale of this fertile land can contribute to destabilizing the local economy, as these farmers will either have to retrain or buy new land further away from their homes due to the lack of available, cultivable land nearby.

Risk zoning is also added to this already complex equation. The lands that are most attractive to the communities to be relocated are partly located in risk-prone areas since they are located between two rivers (blue lines crossing the municipality in the center, and the inter-communality to the left), prohibited to new constructions (yellow square). The plots of land considered for government resettlement sites are therefore limited to the northeast quarter of the catchment area, and to the east of the group of municipalities, where the economic stakes are lower. Households must agree to move to a new location that is less attractive because it has fewer resources. The households that have left often maintain strong ties with their village of origin and go there for celebrations (weddings, religious festivals, for example). Others will also leave, but will choose self-construction or reconstruction to limit the distance from their sources of income, the services and infrastructure of the city and their villages of origin (purple oval). Other families will decide to rebuild on site by taking a few preventive measures such as moving back from the river and reinforcing the main living area (red crosses on the figure). Finally, some households leave permanently, joining family members or relations further away in the administrative territory of the province, or sometimes even on another island of the archipelago (Moatty 2015).

---

5. The word *lahar* is an Indonesian term that can be defined as a hyper-concentrated flow of water and volcanic deposits (Lavigne and Thouret 2000).

After a disaster, the choices in terms of territorial restructuring are marked by the work of memory around the notion of common heritage. The notion of heritage and its symbolic character plays a major role in the success or failure of relocation policies. Di Méo (1998) describes two types of heritage values, which he refers to as community cement: material realities (e.g. a building, a memorial) and abstract phenomena (collective memory) which are related to the former, but which can be moved and recreated elsewhere, differently. As an example, the inhabitants of Bacalor (Philippines) who were relocated following the Pinatubo *lahars*, go to the church of San Guillermo every Sunday. This church, which is a symbol of the disaster, has even become a tourist resource (Gaillard 2002) even though the community has relocated elsewhere. The habit of the inhabitants of the relocation villages on Merapi to go to the "disaster villages" for the great moments of collective life responds to the same logic of attachment to place.

5.3.2. *On-site reconstruction, in the continuity of pre-existing structures*

Despite destructions, disasters do not wipe out the past and reconstruction does not start from a blank page on which everything would be possible. Reconstruction is also a matter of continuity and persistence of pre-existing trends and models, inherited from planning policies and ways of living in the territories. Therefore, on-site reconstruction as close as possible to the ex-ante state can also be explained by the geographical inertia of existing models of land ownership, economic activities and social relations (Alexander 2010). This form of reconstruction can then be considered as a basic form of resilience because it is less costly (no study or mitigation work, or compensation for displaced persons) and faster. By being part of the continuity, it does not require the reorganization of activities and it ensures the maintenance of the former social fabric. Rebuilding on-site is not necessarily synonymous with rebuilding to the same standard, although experience shows that they are often correlated (Moatty 2015). French examples of on-site reconstruction are numerous. Compared to what is done in other countries, the relocation of houses in France is relatively less numerous and the notion of "right to rebuild" helps to explain this. "The right for an owner to identically rebuild a destroyed or demolished building was recognized by Law No. 2000-1208 of 12/13/2000 relating to solidarity and urban renewal (SRU law)" (Article L. 111-3 of the town planning code[6]).

---

6. See: https://www.legifrance.gouv.fr/codes/article_lc/LEGIARTI000031210179/.

Four cumulative conditions allow the right to rebuild (in addition to the obligation to apply for a new building permit), the building must: (i) have been damaged less than 10 years ago, (ii) have been built in accordance with a town planning permit[7], (iii) not be affected by a provision of the Local Town Planning Plan, of the local map or of the Risk Prevention Plan that explicitly hinders rebuilding and (iv) not represent a danger to the life of the occupants. In practice, Article R-111-2 of the urban planning code stipulates that the right to rebuild is cancelled in the case where "the occupants would be exposed to a certain and foreseeable risk likely to seriously endanger their safety". In this respect, properties damaged by torrential floods cannot claim the right to reconstruction.

5.3.3. *Post-disaster as a catalyst for inequality?*

Post-disaster recovery is not a neutral process; geographic and/or sectoral prioritization is necessarily carried out in the programming exercise for the distribution of aid. Consequently, inequalities materialize through local variations in temporalities that can be observed between individuals, communities and portions of the territory (see Figure 5.1). The post-disaster period can represent a risk of impoverishment of already marginalized and precarious populations (Hallegatte et al. 2010). On the contrary, it can also be a "window of opportunity" in terms of prevention, but also in terms of reducing economic vulnerabilities, through the redistribution of wealth by restoring inequalities or injustices, and/or by investing in the diversification of livelihoods. The widening of inequalities and injustices in post-disaster situations is strongly linked to the policy choices that are made and the capacities of governments to manage and cope with crises, the solvency of states, the "social capital" of individuals and communities, the accessibility of sites, as well as the political regime in place and the place accorded to citizen inclusion.

As we have seen previously, aid exists to support individuals' recovery. In order to access state aid, in France in particular, individuals must prove that they are in compliance with the law, that is, have proper papers for foreigners residing on French soil, for example, or have a work contract, or a building permit and a title deed for the plot. However, the most precarious populations present more illegal situations concerning their housing (squatted land or land occupied without a legal

7. The legality of the property is assessed on the day of the construction, which means that if a demolished or destroyed property has become illegal, by revision of the PLU or other urban planning document, the owner still has the right to rebuild their property because it was legal on the day of its construction.

title, for example) and/or their work (workers in the informal sector, or in the underground, convivial or family economy, for example). Moreover, these populations, who avoid attracting attention, will not necessarily claim the aid to which they are entitled (Moatty et al. 2019). The increase in prices in the post-disaster period is also a vector for increased inequality in that the inflation observed almost systematically after a crisis widens the gaps between those who have access to savings and can mobilize them to cope with the increase in prices, and those who do not and who cannot take out credit either. Land use policies such as the creation of buffer zones or no-build zones are also often unfavorable to the most vulnerable (Ingram et al. 2006). In the case of coastal areas, the establishment of a no-build area on the coast takes people away from their sources of income and livelihoods, as well as from their network of acquaintances. Experience shows that these economic and social pressures on households often lead them to return to "dangerous areas" in terms of natural hazards, which are less dangerous in daily life than being deprived of access to vital resources (Boen 2006).

Media coverage of disasters prioritizes certain areas of reconstruction, selected based on cultural, social, political and technical biases that shape how the public perceives risk (Miles and Morse 2007). The mass media present only part of the disaster and highlight unusual and sensational facts to gain ratings (McEntire 2012). Thus, in their study on Banda Aceh, Vinet et al. (2011) show that local variations in the temporality of reconstruction are due in part to the central or peripheral nature of the territories. Reconstruction seems to be faster and better organized in central areas. Indeed, in the case of Aceh, while 190% of financial needs are covered in the city of Banda Aceh, only 42% are covered in the southern district of Aceh Barat Daya (Vinet et al. 2011).

Faced with these inequalities, and in addition to the impetus of solidarity, individuals and communities mobilize their "social capital" to fill the gaps in collective action (Tipple 2005; Bourdieu 2006; Hernandez 2009). This "social capital" is defined by Bourdieu (2006) as

> the set of actual or potential resources that are linked to the possession of a durable network of more or less institutionalized relations of inter-knowledge and inter-recognition; or, in other words, to belonging to a group as a set of agents who are not only endowed with common properties (likely to be perceived by the observer, by others or by themselves), but are also united by permanent and useful links.

The work of Fleming and Ledogar (2008) complements this definition by stating that, as a "characteristic of communities, [social capital] includes attributes such as trust, reciprocity, collective action, and participation". For households with limited "social capital", the recovery pathway will be more circuitous and longer, further widening inequalities and the gap with other members of society (Oliver-Smith 1991; Gaillard 2008; Hernandez 2009; Oliver-Smith 2020). Thinking about capacities therefore means trying to identify the reinforcing and balancing links between resources (available and effectively mobilized) and vulnerabilities in the face of a given situation in order to respond to it.

5.4. Conclusion

In spatial and temporal terms, post-disaster recovery is neither uniform nor homogeneous, and there are local variations in temporality from the individual scale to that of neighborhoods and social groups. These variations have their origins in the severity of the damage, the location of the affected households and their isolation, the reconstruction strategy chosen (identical or different, on site or by relocating), and the structuring socio-economic factors (economic means, access to resources, knowledge in key areas).

At the beginning of the recovery process, the main objective of financing is to provide liquidity quickly to ensure that the crisis is addressed without delay (not allowing situations to drag on, and avoiding post-crisis precariousness becoming everyday precariousness in the long term), and then, as time goes by, the objective is to succeed in mobilizing sufficient resources to rebuild the territory. Regarding the objective of implementing the *BBB*, the adaptation strategies that can be developed during the post-disaster period are not all aimed at reducing risks, but can also be oriented towards upgrading the territory's infrastructure, reducing substandard housing or even reducing energy consumption. They thus contribute to reducing the root causes of the population's vulnerability.

Faced with this challenge and the need to make decisions quickly, governance has been adapted and restructured at the interdepartmental level. Nevertheless, with the exception of a few specific examples of real integration of local actors and populations, the approach remains *top-down*. Often, the integration of the local stratum of the socio-territorial system is limited to simple consultation. And, despite a sometimes clearly stated desire to rebuild elsewhere and differently, pre-existing operating methods – and the associated structural vulnerabilities – prevail and cause "voluntary" reconstruction strategies to fail. The needs of disaster victims, first and foremost the need to act and decide quickly in a context of multiple uncertainties, combined with economic, political and territorial constraints, also limit the scope of possibilities. Therefore, despite strong political decisions, changes are made at the margins and existing structures persist. Finally, exploring the possibilities of transforming the recovery period into a “catalyst” for prevention operations leads to the question of how to better prepare for the post-disaster period, in order to reduce the disastrous effects of decisions made hurriedly, and which plague territories and societies for decades.

6

Crises and Territories: Legacies, Inertia and Dynamics

Nancy MESCHINET DE RICHEMOND[1] and Yvette VEYRET[2]

[1] *Université de Montpellier III, France*

[2] *Université Paris Ouest Nanterre, France*

The territories affected by crises or natural or technological disasters can be seen as heirs to a past that is sometimes poorly understood and that depends on old (political or economic) choices that need to be placed in the context of a given period. For example, the establishment of industries in the 19th and early 20th centuries was a response to technical imperatives (proximity to water, access to raw materials or traffic routes, etc.), which are now outdated, but which have led to a certain number of sites that are still active and potentially dangerous for the populations living in the vicinity (e.g. the chemical corridor to the south of Lyon). More precisely, territories and crises constitute objects of study that cannot be understood without a contextualization that makes it possible, on the one hand, to reconsider the risk/territory articulation (with or against (Meschinet de Richemond et al. 2010); one crisis or several (Peinturier 2020); simple links or complex dynamic constructions) and, on the other hand, to question the temporalities of the links between crisis(es) and territory (in other words, their dynamics among "inertia", "durability" and "resilience"). This contextualization is necessary to grasp the divergences, the telescoping between the spatio-temporal scales of territorial evolution (e.g. the development of the Mediterranean coastline within the context of the 30 glorious years in France following the end of World War II, or urban sprawl at the end of the 20th century) and the spatio-temporal scales of crises (multiple

Territorial Crisis Management,
coordinated by Richard LAGANIER and Yyvette VEYRET. © ISTE Ltd 2022.

scales depending on whether we consider the immediate context of the crisis or its deeper context, the root causes of the PAR model – pressure and release – of Blaikie et al. 1994). These questions of scale are central which lead, depending on the choice made, to fit into different paradigms (Quenault 2013, 2014, 2016; Reghezza-Zitt and Rufat 2015). The paradigm presented by Quenault is based on the acceptance of disaster as an inescapable event in a context of global changes involving a systemic consideration of complexity. It follows the "risk paradigm focused on physical phenomena [...] and [the] hazard paradigm focused on vulnerability", both biophysical and social (Quenault 2016, p. 56). These different approaches are closely linked to evolutions, successive postmodern shifts, in the way of considering "Man/Nature" relations (Quenault 2016, p. 55) and therefore also the relations between humans and space, territories and the process of territorialization as a whole. The complexity also arises from the fact that we are faced with a tiling rather than a strict succession (Meschinet de Richemond 2016) of these paradigms. In particular, we can ask ourselves in what way a crisis (or crises) reveals the diversity of relationships to territories and in what way the manner in which it (or they) are understood and managed bears witness to the perceptions/representations at work in the (re)construction of contemporary societies' relationship to the world. Beyond perceptions, there are many forms of inertia or brakes observed during the course of a crisis. They touch on cognitive, behavioral, organizational and functional aspects that need to be explored.

6.1. Crises and territories: always complex articulations, in constant renewal

6.1.1. *Crises and territories: scales, actors and contexts*

Considering the relationship between crises and territories is difficult insofar as the complexities of the two themes are intertwined and the visions of the various actors involved do not always overlap.

Thus, crises correspond in some cases at least to a rupture, which can threaten the previous social order insofar as this traumatic fracture, which is difficult to repair, is often accompanied by the loss of the reference points that underpin daily life. A crisis disrupts or destroys the habits of life, the way individuals think about themselves, the way they perceive space, their way of living. In this respect, temporary relocation far from the area of daily life affected by the disaster, while a solution for preserving human lives in the short term, can be a factor of social over-disaster (or co-disaster) insofar as it does not allow for the maintenance of relational links (family, social, cultural or ritual) in the long term (Gaillard 2006, 2010) and does not take into account the importance of these relational links in terms of resilience. The failure to take into account these micro-territorialities of everyday life in the "habitus" of techno-centric crisis management not only deprives this management of important levers of action but can lead to obvious failures that have been regularly deplored since the 1970s; the work of Quarantelli (1984) or Blaikie et al. (1994). The shock of a crisis and the emotional intensity amplified by its media treatment too often lead to a reduction (in the sense of Chouquer (2008)) of the consequences of the event to the materiality of the destruction, obscuring the complexity of the territorialization processes themselves disrupted by the event. These "top-down" management methods finally demonstrate, for many crisis management and reconstruction actors, the power of cultural representations of the territory as a support space, a static physical object, and not as a dynamic system of interactions between populations and their living spaces, in all their historical and social depth. For example, the emergency rehousing actions and assistance implemented by developed countries and/or international organizations in Indonesia following the devastating tsunami of December 26, 2004 (Vinet et al. 2011) highlight the difficulties in moving away from a technocratic, top-down and one-time approach, which struggles to respond to the real needs of the victims because it does not include them as full-fledged actors in the process. Rehousing programs can be inadequate because the choice of architecture does not take into account the cultural, social or religious context: a separate kitchen will be desired if there is no mixing of men and women for the preparation and eating of meals, contrary to the dominant culture in the countries of origin of international aid where an open kitchen will be appreciated. Finally: "The inadequacy of the rehousing programs put in place following the December 26, 2004 disaster in Aceh [Indonesia] leads to a dramatic deterioration of the victims' living conditions" (Vinet et al. 2011), a statement that can be articulated with Augustin Berque's reflections questioning the link between architecture and originality of place (Berque 2018). Crisis or post-crisis management that is poorly adapted to local particularities is not only ineffective; it also reduces the society's capacity for recovery (resilience, as some actors would say) and hinders the advent of satisfactory territorial functioning (we will not speak of a "return to normal" insofar as the situation prior to the disaster is rarely optimal in many so-called Southern countries). Such situations illustrate the fact that inappropriate crisis management contributes to territorial inertia by creating or perpetuating dysfunctional situations. The first inertia is perhaps that of continuing to reduce the territory to a support space by minimizing (less in discourse than in action) its multiple complex dimensions (Di Meo 1998; Bonnet 2012).

Another complexity is that reactions to the disaster depend on a large number of factors, both immediate or short term (personal history) and long term (national history or the history of the social or family group). These reactions may depend on the history of the populations concerned, their relationship to the territory, their personal relationships, the existing networks of friends, family and mutual aid, as well as on the previous crises experienced by actors, the memory of which interacts with the new crisis (Benitez 2019; Peinturier 2020). Peinturier thus shows how much the personal and family experience of hurricane crises prior to Katrina (which occurred in 2005) influences the desire or decision (and the conditions under which it is made) to stay, evacuate or return for the inhabitants of New Orleans or Louisiana.

The dynamic dimension of the relationship to the territory of risk and crisis has varied over time and varies according to culture. We can underline the real reversal underway (and begun in the last third of the 20th century) which sees the ever-increasing promotion of the ephemeral (programmed obsolescence, fashions and disposable products within the framework of a consumer society), of short time and immediacy in connection with the rise of NICTs (increasingly portable computers, the rise of the Internet and resources accessible online, social networks), despite attempts to slow down the process (sustainable development, awareness of environmental changes). This reversal is linked to important territorial changes in connection with the processes of urbanization, globalization and the coastal/maritimization of people and activities. In a context of profound change, the meanings of "security" and "crisis" are changing. While the urban population reached 30% in 1950, 55% of the world's population (4.2 billion people) were residing in urban areas in 2018 and this figure is expected to reach 68% by 2050. (UN-DESA 2019). Even if the contrasts between continents or countries are strong, the United Nations report emphasizes "sustainable urbanization" as the main lever for development to enable all citizens to live in a "safe environment". Awareness of the complexity of the interactions of human societies with natural systems (including, but not limited to, climate) has led to an emphasis on the global nature of dynamics that are potential sources of crisis. Thus, the second National Plan for Adaptation to Climate Change (PNACC2 2018–2022) explicitly aims to "better protect the French against extreme climatic events, as well as to build the resilience of the main sectors of the economy in the face of climate change" (quote from F. de Rugy, Minister of Ecological Transition and Solidarity) by playing on the binomial "mitigation/adaptation" and by emphasizing consultation and cooperation between actors. Even though the relationships between extreme hazards and environmental change are particularly complex and contrasting, the ministerial communication highlights the link between the two in a global perspective, with the PNACC2 being part of the European strategy for adaptation to climate change (2013) and the Paris Agreement (December 2015, COP21). Similarly, "smart cities" are more in line with a dynamic approach than they embody a new urban model as such: the fight against risks and crises is part of a global approach, integrating new technologies in the service of a management that is intended to be more sustainable (but paradoxically short term). For example, the city of Angers committed "at the end of 2019 to an intelligent territory, that is, equipped with technological and digital tools for efficient management of resources and the creation of new, more virtuous services for users [to] promote the acceleration" of the ecological transition and sustainable development. The aim is to "develop the territory and accompany the inhabitants towards a new economic, social and environmental model"[1]. The actions are declined up to the individual level (even if the calendar of events is disrupted by the Covid-19 crisis) and, through the ecological transition conference, encouraging citizen participation thanks to communication tools such as the Internet or social networks. The *Angers Loire Métropole* website states that the aim is to "optimize urban services through the use of technology in order to generate savings for the community while accelerating the ecological transition: this is what the intelligent territory will allow, applied to the fields of energy and housing, the environment, mobility and security"[2]. But if this "major shift" is described as "historic" by the elected officials of *Angers Loire Métropole*, it is nonetheless possible to wonder: can the smart city approach really reconcile integration into globalization and sustainability while ensuring the "security" of all, this "security" including the prevention of natural and industrial crises?

The emergence of this global dimension in the relationship to the territory is therefore recent, and does not replace a more fragmented and local relationship (in connection with the space experienced, appropriated by individuals or social groups over time). Indeed, while in the 18th and 19th centuries, the countryside was associated with stability (the eternal order of the fields emphasized by Rousseau), with a space where risks, particularly societal risks, seemed to be excluded, the city appeared to be a place of risk, a place of potential catastrophe. The history of cities is punctuated by the great disasters that have marked them, including sometimes in their materiality and their urban structure, such as Lisbon, whose reconstruction after the tsunamigenic earthquake of November 1, 1755 is an example of urban planning that today constitutes an important tourist resource for the city. During the past centuries in Europe, wars, great epidemics (plague, cholera, etc.), fires, earthquakes, floods, fueled the imagination of survivors (macabre dances in the Middle Ages) and strongly marked the collective unconscious for which the city is a danger zone. But the ambivalence of the representations attached to the city (city of refuge or city "archipelago of danger"; Reghezza-Zitt (2005)) is growing: the visibility and media coverage of the brutal and specific disasters that have affected

1. http://www.angers.fr/l-action-municipale/transition-ecologique-et-developpement-durable/index.html.

2. http://www.angersloiremetropole.fr/medias/62761-l-avenir-est-au-territoire-intelligent/index.html.

cities in recent decades often outweigh the benefits and advantages of the city built over longer periods of time, and which still fuel the rural exodus in the countries of the South, where the city offers better access to healthcare, employment and education.

Moreover, numerous forms of feedback show that the disaster is sometimes poorly understood, and many obstacles appear that explain the poor or insufficient consideration of the event. This consideration is a function of the perception of the process at work, of the reaction of the populations accepting or not the event according to their conception of the world and of society (secular or religious; social mix and mobility more or less strong or valued), their status (citizen, undocumented), their relationship to time and space (primacy of the group or lineage over the individual, sedentary or nomadic society). The relationship to place associated with the relationship between group members (kinship, friends) and therefore to social structures is an important factor determining the reaction of populations faced with the crisis. Economic considerations also come into play, to flee the dangerous space or, on the contrary, to remain there. The organization of territories and political actors, in particular, constitutes an important aspect of effective or ineffective crisis management, depending on the decision-making methods of these actors, their speed of mobilization, their preparation and their organization. These multiple obstacles differ in rich and in developing countries. In the former, there is generally a crisis management organization, sometimes state-run, capable in principle of dealing with the situation. The relatively low number of victims reflects the effectiveness of management, which has improved spectacularly over the last two centuries. But on the scale of the last 30 years, the situation is more contrasted and improvements are often indispensable, as shown by the Japanese examples of Kobe in 1995 and Fukushima in 2011. In France, crisis management could probably be even more efficient by reducing the number of levels of intervention and decision-making, which have multiplied in recent years. In many countries of the South, especially the poorest, the organization of relief and aid to the poorest are often insufficient due to a lack of resources (human, material and financial) and sufficient preparation. In this case, humanitarian aid from the United Nations and numerous non-governmental organizations (NGOs) flows in, necessary and often effective in the short term, but sometimes a source of problems or even tensions in the longer term when it replaces the State and local actors. The example of the Indonesian tsunami mentioned above, or the Haitian earthquake of 2010, clearly illustrates this. The assessment must therefore be qualified in order to improve these interventions, which remain indispensable.

These interlocking, overlapping or successive complexities in the relationships between crises and territories generally result in shifts and maladjustments due to difficulties in adjusting one or another element of the system to the unforeseen events inherent in crisis situations. In rapidly evolving situations, the time it takes to adapt is perceived as detrimental to efficiency, as a slowness or inertia that needs to be corrected to improve management. This inertia is related to several aspects of management, exacerbated by the crisis.

6.1.2. *Obstacles linked to conflicts and the interplay of actors in crisis management*

6.1.2.1. *Numerous and complex systems*

The responses required to deal with the consequences of a crisis are based on territories. Thus, in France, the 2004 law on civil security modified the planning of emergency services according to the scale of the disaster. Plans for the organization of services in crisis situations are now envisaged, and the last decade has seen an increasing desire and affirmation of a territorialized risk management. Since its launch in 2007, the *Plan séisme Antilles* (PSA) (2007–2013) has enabled work to be carried out in schools, vulnerable social housing and the Departmental Fire and Rescue Services. The second phase of the plan defined in consultation with the communities envisaged four main lines of action between 2016 and 2020: reduction of the vulnerability of buildings, support for development and construction actors, public awareness of seismic and tsunami risks and preparation for crisis management, and improving knowledge on hazard, vulnerability and risk.

The National Steering Committee (*Comité de pilotage national*, CPN) meeting in 2019 in Guadeloupe associating the elected officials of the West Indies, the representatives of the State (prefectures, general direction of risk prevention, general direction of Overseas) insisted on four points:

– the realization of a review of the plan during the year 2020 in order to prepare the third phase of the plan;

– seismic retrofitting and prevention, priority of the next European operational programs 2021–2027;

– the support of the project management of communities which must contribute to accelerate the works of paraseismic reinforcement of buildings;

– the development of shared references of techniques and costs of seismic works, intended to optimize the safety of buildings.

At the national level, the second national plan for adaptation to climate change (PNACC 2) for the period 2018–2022, which aims to strengthen the coherence between adaptation and mitigation policies, has been allocated 3.5 billion euros over five years. The evaluation report of the 2011–2015 national adaptation plan showed the need for a national plan that is more closely linked to territorial plans, since the importance of the regional and local levels for the implementation of adaptation actions is no longer in question. The second national plan therefore largely involves local authorities in adapting the territories of metropolitan and overseas France to climate change by 2050. Faced with multiple dangers (flooding, risks of submersion, floods, cyclones, heatwaves, drought, fires, reduced snowfall and its consequences, damage to biodiversity and the economy), the plan focuses on six areas of action: governance and management; knowledge and information; prevention and resilience; adaptation and preservation of the environment; vulnerability of economic sectors; and strengthening international action. The territorial articulation of the adaptation policy must be based on coordination between the territorial and national levels, within the framework of the development of regional plans for planning, sustainable development and territorial equality (*Schémas régionaux d'aménagement, de développement durable et d'égalité des territoires*, SRADDET). The SRADDET reflects a desire to consider a territory (the region) in its globality, complexity and systemic way; nevertheless, the integration of risks and crisis management is a function of regional specificities.

The organization of the civil protection response (*Organisation de la réponse de la sécurité civile*, ORSEC) at the departmental, regional, maritime and overseas levels allows the mobilization and coordination of civil society actors under the unique authority of the prefect (departmental, regional or zone). If the disaster affects several regions of the French territory, it is the national level that is concerned under the direction of the Minister of the Interior. Protection of the communal territory is the responsibility of the mayor, who is the main actor of crisis management at the local level. But adapting territorial management to the multiplicity of actors and vulnerabilities linked to mobility and new urbanization also leads to coordination difficulties, particularly between the State and local authorities.

6.1.2.2. *Obstacles linked to the territorial and decision-making hierarchy and to coordination difficulties*

The multitude of decision-making levels corresponding to as many territories and the large number of associated actors are often obstacles to effective crisis management. In Lima, for example, the organization of crisis management is the responsibility of political decision-makers in their area of competence, but the complexity of the territorial division in the Lima and Callao metropolitan areas, that is, "53 political and administrative entities: 2 regions, 2 provinces and 49 districts form a kaleidoscope of territories" (Metzger et al. 2014), makes crisis management more complex. The different entities that make up the Lima/Callao metropolitan space over a limited area are responsible for a political and administrative fragmentation of the agglomeration that questions the governance of the Peruvian capital, especially since the specific competences of the actors of the different levels of decision-making are poorly defined, particularly with regard to the management of crisis situations. Thus, in the event of a major earthquake affecting the entire urban territory, crisis management is likely to be fragmented between the various political–institutional levels and the multiple political–administrative entities mentioned, which have full competence to decide and intervene on all or part of the territory in the event of a major crisis. Faced with such complexity, the authors emphasize that in order to manage a crisis,

> [...] it is necessary for decision-makers to have a global vision of the needs of the territory, and for them to know the location and characterization of the available resources, particularly for immediate intervention. The relevance and effectiveness of action depend on both the scales of decisions (political–administrative), the knowledge of decision-makers and the location of the decision-making and intervention centers (Metzger et al. 2013, 2014).

In some cases, specific territories to manage the crisis may emerge when disaster occurs, as shown by the example of Haiti, particularly due to the intervention of new actors (NGOs) and the limited response capacity of local authorities. In order to manage the disaster more effectively, the places of power located in the city center were moved to new strategic locations near the Port-au-Prince airport, the arrival point for international actors and logistical aid. Sometimes, during crisis management, two types of territories can coexist, more or less overlapping, which can result in conflicts between management actors. These territorial conflicts also expose social conflicts. For example, Huret (2010) points out that in certain neighborhoods of New Orleans where public authority has ceased to be exercised, reactions of self-defense have arisen on the part of the local population. In the Algiers neighborhood, which was not flooded but designated as an evacuation site by the authorities, a group of residents organized a vigilance committee, blocked access roads and patrolled with firearms.

In a developed country such as Japan, the disaster caused by the 2011 earthquake and its consequences demonstrates insufficient coordination between the State and other administrative levels. This situation also shows an incomplete definition of the attributions and responsibilities of actors at the different levels of intervention. To this must be added insufficient preparation for the crisis. Beyond the dominant position of the State in the management of the disaster, the latter was largely taken over by the municipalities and in a manner adapted to the specificity of the situation (direct effect of the tsunami and/or the effect of the nuclear accident). The municipalities had a preponderant role in the protection of the population. Ahuir

(2016) took up the indications of the mayor of Iwaki to emphasize the importance of management at the communal level, the isolation of certain communes from the prefecture, the difficulty or even the incapacity of the higher levels of decision-making to provide the necessary means for crisis management. A certain inconsistency of successive evacuation orders complicated the task of the crisis manager and showed the uselessness of previously established plans due to their underestimation of risks. Already during the Kobe earthquake and associated fires in 1995, Menoni denounced the “rigidity of the organization of the Japanese system” (Menoni 2001) and the difficult communication between actors. During the disaster, those in charge rigorously implemented previously established procedures, which led to detrimental delays in the transmission of information and the deployment of relief efforts. It was not possible to mobilize the military without a very specific written request from the governor stipulating the number of men, boats and planes needed. In addition, the governor had to produce a precise report of the situation and the affected areas to obtain this military support. The care taken to respond to the crisis and the extreme centralization of Japan did not stand up to the magnitude of the disaster and the need to act quickly.

The observation that there is a lack of resources, both human and material, as well as linked to the unequal experience of elected officials and managers, combined with the objective of achieving economies of scale, has been made by the Japanese model. The drastic reduction in the number of Japanese communes was intended to give the new, much larger and more populous communes a greater capacity to respond. But the regrouping of Japanese communes has resulted in structures that are too large and ultimately unsuited to crisis management: these new communes certainly have greater resources than the old ones, but they lose a certain form of territorial coherence that is necessary for optimal management and probably suffer from the breakdown of links and social networks previously established in smaller territories. In France, the often very small municipal structures cannot have the human, structural and financial capacity to deal with a major crisis. The inter-municipality could be an acceptable response for better crisis management. Going beyond communal boundaries can in many cases optimize the capacity for collective response. Inter-municipality should contribute to the support and assistance of mayors thanks to centralized and efficiently located material resources. The provision of teams trained in crisis management from a center of expertise interfacing with State services but retaining the police power of mayors could facilitate crisis management (Ahuir 2016).

The shortcomings of authorities during the earthquake that killed more than 10,000 people in Mexico City in 1985 explain why civil society took charge of crisis management. The first aid efforts were made by social organizations, and neighborhood associations dealt with the emergency in the very short and medium terms. Thus, popular organizations, independent sectoral organizations, universities and volunteer rescue groups were effective, indispensable and local actors in the face of disorganized federal and local authorities, and a State considered corrupt and incapable (Melé 1998). The capacity for self-organization and solidarity that was expressed was an example for the political life of the country. During the 2017 earthquake, the improvised civil collective action of the inhabitants quickly manifested itself to provide relief in the destroyed areas. The observation made during numerous feedback shows that the disaster victims themselves are the first rescuers and actors of crisis management, especially during the emergency phase (Bourrelier et al. 2000). New communication tools are proving to be valuable tools today: during the 2017 earthquake in Mexico City, the Internet and social networks provided crucial information on crisis management from both civil society and institutions (Valette 2018). But the tools, if they provide technical solutions, do not solve the difficulties of cultural, cognitive and behavioral origin. Bernadette de Vanssay (1999), after her investigations during the volcanic crisis of 1976 in Guadeloupe, the flooding of Nîmes in 1988 or the industrial disaster of Basel in 1986, indicated: "The information of those in charge is often erroneous and fragmentary [...] It is the absence of listening to the witnesses of the crisis situation which, by sending them back to their routines, allows misunderstandings to persist and increases the deficit of understanding between officials and citizens". The paradoxical observation of the permanence of maladjusted behaviors (and identified as such after major events that could have been expected to lead to significant awareness and decision-making) is still relevant and must be linked to the different practices of space according to the actors, to the multiplicity of ways of living and to the complexity of contemporary processes of territorialization that articulate global and local scales.

6.1.2.3. *Information and rumors*

Numerous studies have highlighted the fragmentation of actors' representations in space and time, which makes it difficult to take them into account in crisis management, even though this is an important key to its success, as de Vanssay (1999) points out: "Before each risk prevention plan is implemented, a study of the representations of the populations at risk should be undertaken". The absence of internal communication or poor communication (transmission of bad information, minimized facts) is an aggravating factor. Sometimes, as in Japan, information about risks and crises seems to be well disseminated, yet in Fukushima, the dissemination of information was not fast enough, particularly among the actors in the territories who had to make the necessary decisions in the disaster situation (Ahuir 2016).

When the information circulates poorly or is insufficient, the population, which remains in uncertainty, seeks explanations to reduce anxiety. Inaccurate explanations of the event may then emerge, and the invention of messages from bits of information is transmitted and amplified. Rumors appear, revealing an informal void, that often lead to pointing fingers at those responsible, as well as to explaining what was not previously explained (Weiss et al. 2006). The rumor, which often develops a "conspiracy theory", "an unverified account of an event, which circulates in a group in relation to emotional relationships, develops in a general 'climate' which is conducive to it, the two main characteristics of which are: collective emotional tension and the absence of true information" (Mucchielli 2009). Thus, in the spring of 2001 in the north of France, the Somme River burst its banks and flooded its valley over 6,000 hectares for more than two months, affecting nearly 3,000 people (1,300 of whom had to be evacuated from their homes) and damaging 1,500 homes. These floods were perceived by the inhabitants of the Somme, relayed by the media and local elected officials, as a plot hatched in Paris. The waters of the Oise were considered to be diverted towards the Somme in order to avoid the flooding of the Oise valley and "to protect the Parisians" ... even if the confluence of the Oise–Seine is located downstream from the capital. This rumor, known as the "Abbeville rumor", reflects an opposition between Paris, the Ile de France and Picardy, between the center and a periphery, between different political currents. Such rumors have harmful consequences and leave lasting trauma in the most fragile people. In general, in the event of a crisis, despite considerable efforts made to present and explain the situation both by the State and by scientific organizations, the population is still strongly influenced by the media, and in particular now by social networks and the Internet, which are becoming increasingly important to the public and which often contribute to the spread of rumors (Mas 2012). The phenomenon is now taking on a new dimension with the amplification allowed by these social networks and the dissemination of "fake news". Although the media are now developing more and more programs to decode the "truth from falsehood", the motives for the dissemination of these rumors, which are largely emotional, are not or are hardly modified by a communication that mobilizes rational motives. The current debates concerning the risks linked to the Covid-19 vaccine are a new example, where fear of the risk of thrombosis has led some people in France to avoid the Astra Zeneca vaccine without knowing the higher statistical risks of thrombosis linked to other situations (travel) or frequent medical treatments in daily life. To avoid or reduce the spread of rumors, it is essential to have detailed knowledge of the representations and interests of the various actors who act and live in a territory.

6.1.2.4. *Shortcomings of crisis management*

The violence or speed of the event sometimes explains why emergency procedures previously envisaged are not operational, especially if these procedures for preparing for the crisis and warning were not supported by preparatory exercises. In some cases, the absence of a crisis management plan or even of a crisis cell, the insufficiency or lack of warning systems are all limits to acceptable management. Such situations are responsible for irrational and inappropriate decisions and behavior. Thus, the management of the acute phase of the crisis can be out of step with the specificities of the disaster, especially with its magnitude, which makes the envisaged solutions less operational. As for the post-crisis phase, the absence of feedback or the failure to take into account recommendations made by feedback is a hindrance to the effective management of future crises. The Sequana exercise, carried out in 2016 by the General Secretariat of the Paris Defense and Security Zone (*Secrétariat général de la zone de défense et de sécurité*, SGZDS), simulated the course of a major flood (1910-type) in the Paris region. The aim of this exercise was to consider the coordination of crisis management actors, the chain of command and to deploy resources on the ground supported by the civil security forces of European countries (Italy, Spain, Belgium and Czech Republic). A crisis unit was then set up, which revealed the consequences of a number of shortcomings in the management of the Seine flooding, in particular the lack of analysis concerning the evacuation of the populations concerned. Who should be evacuated? How? To which accommodation and how will the return be done? These questions require anticipation, involving in particular the mobilization of reception areas adapted to the specific situations of the populations to be moved (EHPAD population, hospitals, schools, etc.). Should the Italian authorities, in the event of an eruption of Vesuvius, establish regions to which particular categories of people will be directed? The Italian authorities envisage in particular that done by trucks and trains, the displacement of the most threatened populations will be carried out outside Campania thanks to the system of twinning between the Italian regions. This anticipation, however, discussed, seems a necessity that was not considered either during Sequana, or during the flood of 2016. In the latter case, the interventions of the State and communities took place not in anticipation but in reaction (Fujiki 2017). The development of cities and the urban phenomenon in recent decades poses new challenges of anticipation and adjustments to crisis management.

6.1.3. *Crises and urban territories: rupture, bifurcation, sustainability, resilience?*

Does the magnitude of the crises lead to the abandonment of certain sites and cities in particular? Or can we consider that cities hit by catastrophic events still be considered resilient? The answer to these questions is manifold. Campanella (2006), in his study on urban resilience, shows that most cities that have experienced a major event (natural disaster) persist, and few cities have disappeared for such reasons. We should probably distinguish between persistence and resilience. Persistence is not an indicator of the resilience of the urban system. Reconstruction and resilience imply the existence of local networks that promote a return to an acceptable situation, which is not necessarily a return to the previous situation.

After a disaster (earthquake, cyclone), a new situation arises if the city is devastated. This is the decisive choice that the authorities and surviving populations have to make: whether to rebuild in situ or to move elsewhere. Since natural disasters are likely to occur again in the same places, logic would dictate that the city be moved to a less exposed area, yet Lisbon (1755 earthquake), San Francisco (1906 earthquake) or Tokyo (1928 earthquake) were rebuilt in the same place. The same choice was made in the case of smaller cities, such as Messina and Reggio Calabria in Italy, destroyed by the 1908 earthquake that killed 30,000 people. This is also the case of Yungay, a small town in the Callejon de Huaylas in Peru, which on May 31, 1970 was razed by an earthquake and buried under an avalanche from Glacier 511 of the Cordillera Blanca. The disaster caused 18,000 deaths and 200 survivors were counted. Despite their belief that "the avalanche constitutes a catastrophe greater in magnitude even than the eruption of Vesuvius that in 79 AD buried Pompeii", the survivors wanted Yungay to be rebuilt in the same location (Bode 1989). Some cities affected by major crises fail to recover; they are characterized by a total collapse, and the emergency phase is not followed by immediate reconstruction. This is the case of the city of St. Pierre in Martinique, completely destroyed by the volcanic eruption of 1902, which was marked by the collapse of the territorial system (Benitez 2019): the population (28,000 people) was killed by the fiery cloud, the destruction of the harbor and the port paralyzed trade with the metropolis and abroad. A city of lesser size than the one of the beginning of the 20th century exists today and bears the name of Saint-Pierre but the port and urban functions that were the wealth of the destroyed city no longer exist and the main administrations have been transferred to Fort-de-France. Of course, Saint-Pierre still exists in its physical form, but this small city no longer has anything to do with the former economic and cultural capital of the island: its functions and influence are no longer the same. "On the scale of the Martinique territory, we see an inversion of centrality, with the Pierrotine region becoming a periphery of the island" (Benitez 2018). It seems difficult to consider Saint-Pierre as resilient, knowing that the "revival" of Saint-Pierre is relatively recent and follows a period when the commune itself had disappeared (between 1910 and 1923). Benitez (2018) shows how the choice of the territorial and temporal scales of analysis is decisive in considering collapse or resilience, "the case of Saint-Pierre finally corresponds to the archetype of a territorial bifurcation after a paroxysmal shock".

A city that has been largely destroyed can be rebuilt in a very different way from what it was before the crisis. This is the case of Lisbon after the earthquake, fire and tsunami of 1755, which destroyed a large part of the city. This disaster caused a triple rupture: spatial rupture due to the enormous destruction; social rupture, social and family networks were destroyed, and many surviving city dwellers fled the city; temporal rupture, the disaster interrupted daily life, and the time of urgency of reconstruction was different. The response to the emergency was incumbent on the future Marquis of Pombal in charge of the management of the disaster who formulated the priorities to King Dom Jose I: "Bury the dead and take care of the living" (Baudry 2009). Then came the reconstruction, which took place over a long period of time. In 1756, Lisbon was cleared of its ruins and partly rebuilt. In order to anticipate future events, the Marquis of Pombal launched a systematic survey in all the parishes of the country in order to establish a list of the precursors and effects of the earthquake (duration, damage, animal behavior, etc.). The possibility of a new earthquake was incorporated into the plans for the reconstruction of Lisbon, which was freed from its ruins and became a new city with large squares and wide avenues intersecting at right angles. Indications are provided to secure the population in homes whose construction methods are strictly defined to reduce the effects of an earthquake. This recognition of the need to protect the city against the threat of new disasters leads to the introduction of a new construction system, the gaiola pombalina. However, the protection of the inhabitants and the city was not the only motivation of the policies of the disaster. Several other logics guided the choices and negotiations of the post-1755 period, including the limitation of the economic power of the church and the affirmation of a new will to centralize the State. Can Lisbon be considered resilient? Despite the ruptures that the disaster represented, Lisbon was rebuilt, kept its functions as a capital, and moreover this reconstruction was carried out with the idea of reducing risk and possible future crisis. In this sense, Lisbon can be considered resilient, even if the choice of the spatial and temporal scales of analysis must lead to a different characterization of the trajectories of resilience of the large-scale territories and neighborhoods that make up the city. It is therefore important to emphasize that resilience is a dynamic and differentiated process (Benitez 2018), which cannot be considered on its own but always in a systemic way, in reference to a given time and territory.

6.2. Inertia of representations

The insufficient consideration of feedback, the reproduction of inadequate behaviors or actions despite the numerous studies carried out to better understand crises and improve their management (including risk prevention) can only be understood by taking into account the major role of the representations carried, internalized and constructed by the different actors without always being aware of them. The great contribution of social sciences is to show all the complexities of the social, cultural and cognitive ways in which a crisis is taken into account and dealt with within societies and the human groups that make it up, according to different "governmental mechanisms" (Revet and Langumier 2013). To think that it is enough to have accurate and verified information to respond appropriately (provided that material and human resources are available) is a technicalist belief, fed by a tradition of knowledge (re)construction that emphasizes rationality and the disconnect between the researcher and their object of study and elides the situated, affective, conflictual and plural dimension of knowledge construction (Chouquer 2008; Meschinet de Richemond 2016). Crisis management can only be put in place when society, the actors who make it up, have a minimal common representation of this crisis. Representations interact in a systemic way with the other elements of risk and crisis prevention/management; they can facilitate it, make it more effective or hinder it and each crisis feeds and modifies these representations which also depend on the political, cultural, economic and social context. But these representations may differ from one actor to another and the time frames of their development, expression and transmission differ greatly from the time frames of the crisis itself. Numerous studies and surveys in sociology, environmental psychology and geography (de Vanssay 1999; d'Ercole and Rançon 1994, 1999; Glatron 2009; Mas 2012; Peinturier 2020, etc.) show the diversity and complexity of representations and the extent to which major events leave their mark on memories and can influence potential behavior in the event of another crisis. This influence may or may not be beneficial, depending on whether the course of the new crisis is similar to or different from the previous one.

6.2.1. *Inertia of representations on the role of the population and its vulnerabilities*

The exceptional situation represented by a disaster sometimes justifies behaviors that are different from those of everyday life: fleeing from danger under the effect of panic, or on the contrary being unable to flee and remaining on the spot in spite of the danger, as well as compensating for the lack or insufficiency of rescue teams by substituting themselves for them to, for example, rescue the victims (contrary to popular belief, most of the rescues carried out during a crisis or just afterward are done by the victims themselves) (Bourrelier et al. 2000). The intervention of the disaster victims themselves is carried out without any preexisting social distinctions and is characterized by an activity intended simply to provide for the immediate needs of the group (Duclos 1987; Labeur 2008).

An inertia that can be expressed in the event of a crisis is due to the stereotypes and representations of actors regarding the behavior of others involved (populations, companies, managers, etc.): management plans have difficulty integrating new behaviors and certain realities that do not fit in well with inherited but still active frames of thought. Thus, although the need to take into account the population as a full-fledged actor in the management of natural and industrial risks is recognized in France, implementation remains slow. The Seveso 3 directive insists in particular on public access to safety information, participation in the decision-making process and access to justice in accordance with the requirements of the Aarhus Convention. Citizens must have direct access, via the Internet, to information on Seveso installations located near their homes, on accident prevention programs and on emergency measures to better react in case of need. These populations can take legal action if they feel that their rights have not been taken into account when a new Seveso site is installed near their home. The directive also includes provisions to improve the way information is collected and managed. Indeed, such an approach is not totally new, as the provision and sharing of information on hazardous sites has been recognized for over 20 years in France. Local information and consultation committees (*Comités locaux d'information et de concertation*, CLICs) were established in France by the 2003 Risks Act (known as the Bachelot Act), following the 2001 AZF disaster in Toulouse. In addition to CLICs, there are also local information and monitoring committees for waste treatment facilities (*Commissions locales d'information et de surveillance*, CLIS). But the implementation of these committees is slow, held back by a top-down socio-technical culture, a Jacobin heritage or practices that are difficult to question and that consider the citizen less as an active agent of management than as a passive victim who must be rescued. Developments are also not linear and the nature of obstacles changes: the recent technical developments of the "Géorisques" website, which provides the public with abundant data on all types of risks, ultimately lead to complicate the possibilities of information on industrial risks due to reduced functionalities concerning data requests, exports and cross-referencing.

Numerous authors have shown, based on feedback studies, the difficulties and problems posed by crisis management associated with irrational behavior, sometimes involving the impossibility of accepting or even conceiving the event itself. In some cases, when confronted with danger, we note that populations can adopt risky behaviors whose causes are multiple and more complex than they appear, mixing at the same time erroneous representations of risks, individual factors as well as cultural (young men overrepresented, Shabou (2016)) and social (labor market, gendered division of family work, professional constraints). Thus, the human toll of Mediterranean floods in France in recent decades (the Var, June 2010 2011, 2012, 2014, 2018, etc.) testifies to the difficulty in imagining the actual power of water combined with an excessive feeling of safety associated with cars and leading to inappropriate behavior (failure to comply with instructions enjoining people not to move, engagement on flooded sections or fords, displacement of safety barricades, etc.). Aublet (2014) reminds us about the rapid floods in Mediterranean France: "nearly 40% of the deaths recorded during the 20th century would concern drivers swept away by water in their vehicles". Several works on this issue of human losses due to rapid flooding (France, United States) emphasize the interweaving of social and spatial–temporal factors of exposure to danger and the particularly dramatic risk of car-related deaths (Ruin 2007; Shabou 2016; Terti 2017). Despite the results of these works which show the need for a dynamic approach fully integrating social sciences to better anticipate situations of increased vulnerability (and thereby adapt and improve mitigation measures, prevention policies and better adjust the organization of relief to needs), vulnerabilities are still too often apprehended in a static way, as a state or even an intrinsic characteristic of stakes (such as the resistance of a building material) and not as processes, chains of complex causalities. The difficulties of crisis management in integrating the new complexities linked to the dynamics of "systems of vulnerabilities" (d'Ercole et al. 1994) are due to several factors, notably historical, more or less rooted in the behavior and reflexes of actors.

A major difficulty results from the weak interface between the world of research, even applied research, and the world of operational emergency management: tools, methodologies, modes of evaluation and time frames are not the same in these two fields and make exchanges difficult due to the lack of a common culture, of a common operating implicit.

Another difficulty lies in the epistemology and history of knowledge construction. The techno-centric approach still frequently reflects the belief in the illusion of an objective science, with knowledge that is independent of the humans who build it. If there is a consensus today on the need for in-depth studies on hazards in order to build effective risk management (natural or technological hazards), this consensus should not obscure the fact that it was built up over decades of trial and error, sometimes conflicts, and then in-depth research that enabled the constitution of a solid scientific corpus that was not without its uncertainties. Thus, in France in the 17th and 18th centuries (Desailly 1989; Virol 2008), the king's engineers who worked on rivers were above all military engineers: works and structures for protection against floods were considered more in connection with the defense of the kingdom (making fortification systems more effective, routing troops requiring functional roads and bridges, etc.) than in connection with a hydrosystem whose functioning was not yet known, nor in what way it generated floods (Coeur 2003). From the last third of the 18th century, the Ponts et Chaussées engineers were responsible for the development of rivers and gradually built up knowledge of the river system, leading them to move away from a punctual and static approach to development thanks to the observation of the river and its interactions (erosion or embankment) with the work carried out. By questioning the causes of flooding or the degradation of structures, and thanks to technical innovations (cartography, metrology, etc.), the functioning of watersheds is increasingly well known. The technical skills of the Ponts et Chaussées engineers developed in the 19th century and their know-how was recognized, even though major floods (1846, 1866 and especially 1856) showed the inefficiency of raising dikes in the face of extreme events. This accumulation of knowledge over time allowed public authorities to develop prevention with the establishment, in the second half of the century, of the first flood warning systems (1854 in the Seine basin) thanks to the continuous monitoring of rivers (Coeur 2003). This slow and ancient construction of technical knowledge explains the time lag in the dissemination of knowledge, the contributions of human and social sciences being much more recent, including the contributions of a reflective and constructivist approach. The hazard- and techno-centric "risk paradigm" (Quenault 2016) is still largely dominant, even though new approaches centered on vulnerability (and then vulnerabilities) have developed (Chicago School, White 1974; O'Keefe et al. 1976; Blaikie et al. 1994) over the last 50 years. This entry through vulnerability, carried out by geographers and sociologists in particular, is part of the "danger paradigm" (Quenault 2016), which is less widespread than the previous conception.

6.2.2. *Inertia of the representations of the crisis itself*

The limits to effective crisis management are often also related to history, to the existence of social links and networks that maintain ways of thinking that are not adapted to the crisis situation. The culture of the threatened populations is therefore a determining factor (weight of religion, acceptance of danger, etc.). In 1994, in Martinique, Robert d'Ercole and Jean-Philippe Rançon already showed how different the representations of a future eruption of Mount Pelee were between the populations surveyed in Morne Rouge and volcanologists. The vision of the populations remained "fixed" (d'Ercole and Rançon 1994) around the paroxysmal, brutal and brief catastrophe of 1902 (while the other less violent phreatic eruptions were forgotten). Such a representation favors the acceptance of a rapid evacuation in case of an alert given by authorities, as well as induces biases in case of another eruptive scenario, little or not known by the populations, and hinders the construction of a representation of an eruptive crisis lasting several years, a scenario that is nevertheless conceivable according to volcanologists. Thus, the inertia of the representations of the populations, linked to the weight of the 1902 event while the memory of other less traumatic events has faded, will participate in a complex dynamic situation during a future crisis, according to the interactions that will arise from the confrontation between the different components of these representations and the unfolding of the real crisis. Similarly, the way in which previous hurricane crises have been experienced by Louisiana residents explicitly influences the behavior they intend to have when faced with future evacuation directives (Peinturier 2020). Crisis management, like reconstruction, is not only about repairing damage and restoring material goods, but it is also about reconstituting social ties and maintaining or building a collective memory.

In the past, in a predominantly rural and Christian Europe, the events responsible for crises and disasters (floods, earthquakes, plagues, droughts, food shortages or famines, etc.) were not above all brief and brutal events. Historians speak of calamities to designate these long-lasting crises, in particular, agricultural calamities that regularly manifested themselves in poor harvests due, for example, to overly cold springs, overly wet summers, as well as to armed conflicts… Brief and large-scale events are then inserted into a chronology of daily or chronic crises (Meschinet de Richemond 2011). These disasters or calamities mark the life of societies without always being perceived by them as major ruptures. Even if we must be aware of the risks of anachronism and decontextualization (Meschinet de Richemond 2016), numerous and complex aid and support mechanisms, based on experience, bear witness to the society's assumption of responsibility for these crises, and even to their acceptance through interpretations (including religious ones) that make it possible to give them meaning in a daily life that does not separate the political and the spiritual. If material support is often precarious and very partial, functional support is real through a symbolic interpretation giving a status to the event (divine trial, for example) without calling into question the structure of the society. Oral transmission, writings and legends, fulfilled a social function of maintaining vigilance much more than the transmission of reliable information in the scientific sense of the term. Today, Western citizens of the individualistic consumer society have mostly lost this vigilance, even in the face of frequent crises. The great mobility of populations for less than a century also explains the lack of knowledge of the dangers that can exist in the host region, despite the regulations that impose in France to make the risk known to the purchasers of goods or to the tenants, and despite the efforts to raise awareness of the risks that have been widely developed in the last two decades. Added to this is the rupture of the population with nature, the widespread belief in a technical mastery allowing everyone to do what they want when they want, freeing themselves from environmental constraints.

People in some developing countries continue to interpret and understand crises and disasters through the prism of religious conceptions that are rarely taken into account by many international actors. In emerging countries, the influence of communities' traditional beliefs and their way of looking at nature are fundamental in the management of crises that are often perceived as being sent by god(s) or the devil(s). Thus, in the Philippines (Gaillard 2006), Indonesia (Texier 2007) or Mexico (Gavilanes-Ruiz et al. 2009), the various authors report on the very different representations and perceptions of volcanoes that justify the decision of populations to move away from the risk zone or, on the contrary, to remain there. Vulnerability is unequally perceived according to the population's culture. The sector located on the San Andreas Fault in California is densely occupied by well-to-do inhabitants despite the associated risks, but the quality of the landscape largely prevails over the danger and explains the location of the population which, however, knows the risk. The same is true of the landscapes along the banks of the Thames in the London area, which are very attractive despite recurrent flooding. Cognitive dissonance can intervene in the construction of social representations of danger, leading to paradoxes visible in field surveys. Cognitive dissonance leads to the reinterpretation of a situation to eliminate contradictions or difficulties. Thus, Henry (2007), who investigated the perception of flood risk by industrialists in the north of France as part of INERIS' reflections on NaTech risks, confirms the complexity of respondents' perceptions in a beautiful formula: "a paradoxical feeling of being exposed… but safe". Cognitive dissonance appears in particular after traumatic situations; painful or disturbing memories are then distorted, sometimes even obscured or completely forgotten (Mas 2012).

The attachment to the place of living is, in many cases, fundamental to understand that the populations do not want to leave the place of the disaster, to leave would be like a renunciation, an abandonment of a whole part of their life. Few people are ready to make such a sacrifice, even if it seems rational. Some "evacuees" from New Orleans following Hurricane Katrina in 2005 wish to return and refer to themselves as New Orleanians even after several years of relocation to other states (Peinturier 2020). Man's connection to the space they inhabit is very powerful and, as de Vanssay (2010) has noted: "it is of interest to several types of relationships: of identity (allowing one to define oneself), of habitability (living space) and of territoriality (allowing one to define oneself in relation to others)". In short, space is the place where belonging to a community is formed. We must not destroy, "in the name of reconstruction, the tenuous but fundamental ties that people have forged with places and that were highlighted by the earthquake", writes Japanese architect Hitoshi Abe, co-founder of Archi Aid after the Fukushima disaster (cited in (Ahuir 2016)).

The force of habit and family roots can mask the critical reality of the situation. The "culture" associating a territory, ways of acting and representations has long made it possible to explain the way human groups think or act. Thus, disaster management reveals a link between the territory, the identity of the people who live in that territory and the roles that these people have in that context (Revet 2017). The relationship of the individual to the place of living, defined as an identity, historical, relational space, also refers to the sharing of the place by other individuals who recognize each other, are in relationship and can find the traces of a common history (Augé 1994). Attachment to a place of residence allows for the establishment and maintenance of social ties between members of the same community. The network of family and neighborhood relationships plays a major role in the recovery of individuals; thanks to the network, assistance to victims is frequent and rapid, as is help with restoration and cleanup. Family or community ties are an important source of material and psychological support. The existence of networks and the associated attachment to place also explains why some individuals settle in dangerous places (Cannon 2008; Gaillard 2006, 2010): social and cultural ties make it possible to accept natural or technological hazards, and moving to a safer space at the cost of losing these ties appears to be a potentially more serious loss than exposure to the hazard. The relationship with the place of living can also be economic, as Franck Lavigne, Benjamin De Coster, Nancy Juvin, François Flohic, Jean-Christophe Gaillard, Pauline Texier, Julie Morin and Junun Sartohadi have pointed out, showing that the inhabitants of Merapi earn an average of between one and three euros per day, which enables them to live with their families; but if they have to leave their usual place of living in times of crisis, then they have no money to support themselves. Thus, during a crisis, the population has to sell their livestock at low prices, because cattle farmers cannot feed them properly. In addition, keeping the livestock in the village forces the inhabitants to go and take care of the animals (milking in particular) and therefore to return regularly to the village, which is abandoned because it is dangerous. These practices are particularly risky during eruptive periods (Wulan Mei 2013). Thus, if millions of people are settled knowingly in areas exposed to natural risks (floods, earthquakes, tsunami, sectors affected by volcanic eruptions as well as technological dangers of various kinds, etc.), it is an empirical cost/benefit approach at the group level insofar as these areas allow them to live (agriculture on the slopes of certain volcanoes, proximity to the workplace, etc.) in the absence of the possibility of moving or settling in more secure areas.

The relocation of populations in the event of a crisis is often difficult, even though relocation is planned. The earthquakes that affected Italy, notably in 1968 in

Sicily, in 1976 in Friuli and in 1980 in Campania, required the relocation of the affected population but, in each case, they wanted to stay where they were and rebuild in the place where their houses were located before the earthquake. The Italian side of the Alps and the Apennines are home to many villages that have been destroyed at least once by an earthquake and, in the vast majority of cases, have been rebuilt in situ (Signorelli 1992). What is it that drives the inhabitants, who have been severely tested and traumatized, to remain in the very place where they risked death? The answer probably refers to this close relationship with the place, in its identity and relational dimension. This place, a symbol of life before the disaster, undergoes a process of idealization like the disaster. In addition, in some cases, resisting displacement means opposing authorities' decisions, even though they are intended to protect disaster victims from danger. This was the case in Venezuela in 2000–2001 following the Vargas disaster of 1999 when the government decided to move the affected populations to sparsely populated parts of the country. In total, 16,000 families were displaced to 18 of the 23 States in the country. It can be assumed that between 50,000 and 80,000 people left Vargas to move to houses built or recovered by the government. Nevertheless, this policy was a failure, since a large part of the displaced people returned to the coastal region in the years that followed. There are many reasons for this, but the rupture of social ties, particularly in order to find work, was fundamental. In addition, the very mixed reception received by the "displaced" in their new region did not favor their integration, especially since these regions were not prepared to receive this population from Vargas. By the end of 2001, approximately 70% of the displaced had returned to Vargas. Revet (2007) points out that these returns are a sign of the refusal of the government's displacement policy, as well as of the importance for the victims to regain their cultural, economic and friendly references. In many cases, if the place of residence allows the family to be fed, the poorest people remain in dangerous areas, taking the risk of repeated disasters.

In other cases, the disaster may be accompanied by a migration policy, as was the case after the Messina disaster, as in the case of the Po floods in 1951. There, after the disaster, to facilitate evacuation and allow for easier reconstruction, free train tickets and passports were distributed to the population, for what was intended to be a temporary move. The population emigrated to Northern Italy or to other European countries, in some cases permanently. In the longer term, this situation may reduce the ability of these regions and the remaining populations to rebuild and increase their resilience.

6.2.3. *Territorial inertia feeds crises: a territorialization of data unsuited to management?*

Risk is often defined as a conceptual and spatial intersection between hazard(s) and vulnerability(ies), a definition that is widely disseminated today by awareness-raising tools (see ministerial websites[3]), school textbooks and preventive information. But this definition, which has become consensual, slows down the dissemination of research results and feedback on risks or events affecting large urban areas (from Latin American cities to New Orleans or Paris; d'Ercole and Metzger 2009; Hernandez 2010; Reghezza-Zitt 2015; Fujiki 2017). Some urban spaces have the originality of generating vulnerabilities that can be transmitted to the whole territory to which they are functionally connected (and not necessarily spatially): "the vulnerability of a major issue is transmitted to the whole territory by dependence, spatial vulnerability is transmitted to the issues by location" (d'Ercole and Metzger 2009). It can also be transmitted over long distances, in a reticular manner, when the functional links between several distant territories (large metropolises, for example) are stronger than with spatially close areas. November (2002) already mentioned in her thesis the difference between "contiguity" and "connectedness" in the territorialization of risks. The development of GIS has made it possible to make great progress in the spatialization of hazards, stakes, vulnerabilities and damage by specifying indicators such as proximity (distance between stakes and a source of danger; between damages and destructive phenomena) and concentration (density of stakes, sources of danger, types of damages), also making it possible to present maps of gradients of damage or vulnerabilities. However, this progress tends to focus communication with managers and elected officials on spatial vulnerability. Awareness of functional vulnerabilities is slow and still remains in the domain of specialists. The difficulty in representing fragmented and dynamic vulnerabilities, which can link several territorial scales, certainly explains in part the slowness of the diffusion and appropriation of this broadening of the definition. The realization of risk in the form of a crisis does not simplify anything insofar as the territories affected are themselves multiple and changing according to the temporal scales selected. The transmission of vulnerability thus explains why territories far from the initial source of damage (erupting volcano, flooding or cyclonic trajectory) can themselves enter into crisis. For example, six months after Hurricane Katrina, approximately 40,000 students who had left Louisiana were welcomed in Texas institutions. "Still considered to be the 'Katrina Student Population,' and therefore in a temporary situation [...], the State of Texas [must] take special [management] measures: housing, food, supplies, increase in the number of students per class, physical adjustments to infrastructure,

3. https://www.ecologie.gouv.fr/politiques/risques-naturels or https://www.georisques.gouv.fr/.

mobilization of additional personnel. Thus, a territory not directly damaged by a disaster finds itself in a situation where, several months after the event, exceptional measures are taken to ensure the management of a crisis of unknown duration: no one knows how long students will remain in Texas, prevented from studying temporarily in Louisiana institutions, or unable to return to live in decent conditions in their initial living territory" (Peinturier 2020). These territorial entanglements are coupled with the growing territorial complexities already present before the crisis and not always suited to its description and management.

The territorial grids of everyday life do not always facilitate the apprehension of crises. Thus, research work (Cotelle 2014; Peinturier 2020) emphasizes the extent to which the data useful for analyzing and characterizing the crisis following Hurricane Katrina in New Orleans requires an enormous amount of fine-tuned work in building and verifying databases (on post-disaster urban insecurity: Cotelle (2014)), as well as a sorting process that requires, for example, a precise analysis of the vocabulary, whether it is to understand changes in "criminal acts", "insecurity" or mobility. Depending on the organization, the definitions of "home" or "occupied housing" and the methods of calculating "income" may differ to such an extent that reading the changes in the raw figures would lead to opposite conclusions without a detailed critique and contextualization (Peinturier 2020). The available data are structured according to administrative divisions that can be multiple and evolve over time, complicating the analysis and identification of differentiated territorial trajectories on a local scale. For example, the city of New Orleans should not be confused with its metropolitan area, which, under various names, includes eight parishes, bearing in mind that the Saint James Parish was no longer part of it in the early 2000s before returning to it in 2015 (Peinturier 2020). The city and parish of New Orleans overlap almost exactly, but the subdivisions differ according to zipcodes (15) or neighborhoods (72). The latter are subdivided into census tracts and census blocks to aggregate information from official population censuses (Cotelle 2014). However, despite this fairly fine division of the territory, the boundaries of the Security Districts, an important subdivision to better understand the evolution of "security practices" with regard to crime before and after the event, do not always coincide with the boundaries of these census units. Thus, the territorial divisions, which are of course necessary for structuring the data, can at the same time "hinder the analysis" (Cotelle 2014) or the highlighting of dynamics useful for characterizing all the facets of crises and their territorial anchorage. If the construction of territorialized databases is itself complex, the rapid transformations induced by crises also favor distortions between the actors' representations, the image of territories and the reality of the facts and processes at work. Thus, Cotelle (2014) emphasizes that the "rapid functional mutations of territories lead to distortions between perceived insecurity and objective insecurity". The inertia of representations of insecurity and crime, reinforced by media discourse, encourages these distortions. Some actors can also implement territorial marketing, a media campaign to change the image of a neighborhood. These distortions between criminal realities and perceptions/ representations are all the more complex to grasp since real developments can be contrasted on a larger scale than that of the neighborhood, once again questioning the relevance of the administrative territorial division to accurately grasp the functional territorial developments linked to crises. The Leonidas neighborhood is an example: gentrified in its southern and eastern parts after the disaster and impoverished in its northern part, "this neighborhood, which became increasingly safe after Katrina, may thus be the object of positive representations [among] individuals who are aware of the territorial transformations, but it may, on the other hand, retain a negative image for individuals who do not frequent it" (Cotelle 2014). If we are interested in territorial trajectories (degradation/revitalization, resilience), identifying new or emerging dynamics is therefore particularly complicated by the inertia of the territorial divisions used as a framework for data collection and by the multiplication of actors.

6.3. Crises feed territorial inertia

Territorial inertia, which is to be understood not only as the stability of spatial frameworks but as a territorial functioning adapted to daily time, can feed crises in multiple ways. The increasingly complex definitions and ramifications of crises are linked to the increasing complexity of the interrelations between the elements of the daily territorial system, thus multiplying the opportunities for major dysfunctions and the functional transmission of these dysfunctions. The spatial and temporal amplitude of crises can thus be increasingly difficult to circumscribe and specify, even considering the differences of situation and point of view according to actors. This raises the question of the extent to which crises do not in turn feed territorial inertia. Extreme cases concern countries where recurrent and closely spaced disasters, combined with current and inherited political difficulties, lead to a situation where (re)construction never reaches its conclusion. We can also ask ourselves to what extent crises are part of an increasingly globalized system, in which their management, while providing emergency solutions on certain scales, contributes on other scales to the maintenance of vulnerable or ill-adapted organizational frameworks, as well as to the permanence of inequalities and power relations between actors and between countries. Finally, the historical perspective allows us to question the way in which a crisis is constructed, sometimes over the very long term, and remains almost invisible, either because actors are not really aware of this slow construction or because the actions of some of them contribute to this invisibility.

6.3.1. *When crises and emergency become everyday life: the example of Haiti*

How do we "rebuild from the devastating effects of earthquake or refound from the past and history to create the anchors of renewal, making the ruptures necessary to build a just society?" (Buteau et al. 2010). The crisis situation facing Haiti predates the 2010 earthquake, as the Haitian population has experienced the effects of repeated natural disasters in the past: droughts, hurricanes, deadly floods, repeated earthquakes, not to mention the land movements and slope erosion that affect a very mountainous country with very degraded vegetation cover. If we limit ourselves to cyclones, the country was affected between 1960 and 2010 by several particularly violent events: Flora (1963), Cleo (1964), Ines (1966), David (1979), Allen (1980), Gilbert (1988), Georges (1998), Jeanne (2004), not to mention Faye, Gustav and Hanna in 2008, etc. Their recurrence shows not only the impossibility of considering them as isolated events but also how much the consequences of their passage are, in fact, part of the daily life of Haitians. Over time, Haiti has suffered multiple successive or synchronous shocks: they concern health (epidemics of malaria, typhoid, cholera, etc.), they are economic (devaluation of the gourde, price increases, effects of international economic shocks), and political (coup d'état, riots, conflicts, associated with corruption). Crisis situations are associated with these shocks: impossibility to find a job, to have a decent income, to have access to essential health care, to be able to send children to school, etc. Independent since 1804, the country has experienced chronic political instability and is today the only country in the Americas classified as a least developed country: its HDI of 0.532 ranked 149th out of 182 countries according to the UNDP in 2009 and in 2019; Haiti was ranked 170th out of 189 countries with an HDI of 0.510 (Banque Mondiale 2020). The trauma of January 12, 2010 must therefore be seen as part of a multi-trauma system. Previous traumas mark the experience of individuals (Benitez 2018) even if the major earthquake of 2010 tends to overshadow them on an international scale insofar as it caused 230,000 dead or missing, led to a situation of increased precariousness, with approximately 1,500,000 people living in makeshift shelters. In addition, 600,000 people were displaced to provincial cities, losing their family and geographical references (Rainhorn 2012). Despite the international repercussions of the event and numerous emergency aid, Haiti has not recovered since 2010 and exists in a precarious situation aggravated by other natural (Hurricane Matthew in 2016 or Irma in 2017), political and health events that have occurred over the past 10 years. Thus, the year 2020 marked by the Covid-19 pandemic, the passage of tropical storm Laura and socio-political events, recorded many acts of violence. These events contributed to the worsening of the country's economic situation, reducing the purchasing power of a large part of the population and increasing the vulnerability of many Haitians. At the end of 2020, 69% of households reported a decrease in income. Half of them estimate this decline to be more than 60% (Nations Unies Haïti 2021). Many households cannot cope with the various shocks mentioned, their repetition and their consequences. This extreme fragility of Haiti, which characterizes the daily life of the majority of people, reduces the possibilities of resilience for both the population and the decision-makers. In these conditions, humanitarian emergency remains, to face the Covid-19 crisis, to contribute to the feeding of children in particular, to education, etc. Haiti remains an assisted country where misery is even greater than before 2010. The rapid succession of multiple crises finally builds a state of permanent crisis, which prevents a real urban and territorial structuring, especially in Port-au-Prince (Milian and Tamru 2018). This situation itself appears to be an aggravating factor for future crises, insofar as it feeds a high degree of vulnerabilities (social, economic, political, organizational) and does not allow for the development of an overall resilience based on the high resilience of individuals confronted on a daily basis with this chronic crisis situation.

In such a context, the definition of a "normal" situation, or a "return to normal", no longer has much meaning, except to underline the difference in the relationship to the norm depending on the context and actors involved. Beyond the satisfaction of vital needs to ensure the survival of a population, the reality designated through the expression "normal situation" differs according to the cognitive, cultural, political, regulatory or legal reference points of the different actors present, making it particularly difficult to construct a common reference. These differences are themselves factors of vulnerability that persist over time and fuel vulnerability, the breeding ground for future crises.

6.3.2. *Instrumentalization and resilience: the root causes of crises are still present*

Repeated crises that hinder the development of certain countries with limited resources, such as Haiti, contribute to a territorial inertia marked by poverty, political instability and a lack of means to reduce vulnerabilities in view of future crises. It is clear that, for many decades, foreign aid has not been able to resolve these difficulties.

Even when the triggering event is of natural origin, poverty is a major factor in aggravating the consequences of the event in both rich and developing countries. It feeds the crisis and favors territorial inertia by hindering reconstruction, making crisis management more complex, partial and less effective. In many cases (such as Haiti, already mentioned), the poorest populations are located in sectors that are overexposed to danger: proximity to waste storage sites, steep slopes that encourage flooding or landslides, and former industrial sites that have not been rehabilitated. This situation is denounced in the name of environmental and social justice, but these exposed areas abandoned by the more affluent populations, sometimes not registered, not buildable or with a very low land price, remain attractive for the poorest or migrants without official status. Poverty can lead to environmental degradation that can trigger or aggravate the effect of dangerous processes (e.g. landslides favored by the deforestation of unstable slopes), and the disaster often deprives the poorest of their means of subsistence, especially since they do not have access to an insurance system or to credit reserved for wealthier populations. Poverty and "underdevelopment" encourage the use of materials and construction techniques that are not adapted to dangers, and the absence of a transport infrastructure does not allow rescue operations to reach disaster sites quickly. Poverty is linked to an informal habitat that is often fragile, overexposing the inhabitants to violent winds, cyclones and earthquakes, and preventing them from evacuating easily in the event of an alert due to the lack of a working vehicle. While the role of education is systematically emphasized by researchers as a means of facilitating a more appropriate response to disasters as well as easier reconstruction, poverty complicates access to information or safety instructions, either directly (lack of radio or effective communication tools) or indirectly due to a lack of sufficient and widespread schooling, including for girls. During the disaster in New Orleans, 30% of illiterate people (estimated by the NGO, Teach of America) were unable to respond to the authorities' requests regarding their situation and were unable to participate actively in a reconstruction process that required them to become familiar with administrative and legal documents (Hernandez 2009).

Similarly, the issue of land tenure and land rights plays a decisive role in the recovery of populations after a crisis (conflict, natural crisis). For example, without clearly established rights to a given piece of land, programs to rebuild infrastructure or resettle displaced populations will be a source of conflict and will be delayed (USAID, Land Tenure and Property Portal – Land Tenure and Disasters, October 2014).

The so-called radical critical approach emphasizes the extent to which international aid (from states and NGOs), large-scale actions and the resources mobilized during major crises constitute emergency aid but rarely long-term aid. International mobilization can be considerable: a few days after the Haitian earthquake in 2010, donations to NGOs (the Red Cross, Doctors of the World, Partners in Health, Catholic Relief Services, etc.) were already twice as high as those given during the Indonesian tsunami in 2004, which were considered particularly high at the time. Although this external aid is multiform and important, it is nevertheless highly debated, as the Indonesian and Haitian examples show. The proliferation of NGOs in Haiti in 2010 gave the impression that foreign governments and relief workers were trying to put the country under their guardianship, whether real or not, which could not contribute to a well-managed crisis. The assessment made several years after the earthquake points to the waste of aid, its lack of effectiveness and also the competition between the various NGOs, which tended to replace the State. Some NGOs set up emergency hospitals in Haiti in 2010 to treat the poorest people, leading to the disappearance of certain establishments that existed before the earthquake, which lost their place and function. Canvas villages were set up far from the living quarters of the populations who nevertheless continued to frequent them. The large number of NGOs present on the site of the disaster would have required real coordination between them, both in the emergency and post-crisis periods. The UN representative in Haiti sought to preserve the "illusion of governance" by the Haitian State but in practice the State was placed under the control of the UN and "relief" states. The perverse effects of certain aid projects deployed without prior concern for the social fabric and local structures, and for the relationship of populations to their living space, are regularly highlighted by feedback reports, but still persist. The crises that mobilize these NGOs and aid donor states paradoxically reinforce a system that gives more media coverage to their emergency action than to the complex and sometimes perverse effects of their long-term actions.

These emergency aids do not address the root causes, the economic, political, and land systems or the unequal relations between states that persist and maintain the construction of particularly vulnerable territorial systems over the long term. The success and rise of the notion of resilience (Reghezza-Zitt and Rufat 2015), due to its polysemy and the "shift from a risk society to an uncertainty society" (Quenault 2015; Reghezza-Zitt 2015), thus appear to be a way of bringing to the fore the role of populations, local actors and structures. The notion of resilience covers in most cases the idea that the post-disaster period should allow for a better adaptation of populations to the occurrence of future crises; it is underpinned by the idea of improvement, of progress. But what about the populations that do not achieve this improvement? Doesn't resilience lead to a distinction between those who succeed more or less well in overcoming crises and those who do not, thus introducing a kind of hierarchy that can lead to moral judgments, obscuring in the process the long-term role of States or small-scale structures. This notion of resilience, which often insists on the need for populations and citizens to take charge of their own lives during the course of a crisis, can then be analyzed as a way of absolving States of their responsibilities. It can also be seen as a way to avoid questioning more deeply the links between a global economic and consumerist system and crises, including global crises with multiple consequences, which are increasingly difficult to absorb by the system. This injunction to develop resilience (from individuals to companies) does not compensate for the increase in vulnerabilities linked to the rise in competition between territories (in France, "country", "terroir" and "appellation d'origine" labels, territorial marketing and competition between regional "metropolises", etc.), in parallel with the disengagement of the State (decentralization, deconcentration of public policies) and the transfer of responsibilities (without always having sufficient financial means) to more local levels.

We can sometimes observe an instrumentalization of crises, masked or justified by resilience. For example, the uneven reconstruction of New Orleans has led to significant urban reconfiguration, transforming its image (or that of certain neighborhoods) and making it more difficult, if not impossible, for populations considered undesirable or less dynamic for the city to return. The low level of public investment in certain hard-hit neighborhoods after the disaster does not encourage their inhabitants to return and then justifies priority investment in areas where returns are greater. This vicious circle encourages further urban fragmentation. The central business district, which was spared, benefited fully from tax credits granted by the federal government for the rehabilitation of historic buildings. Private outside investors seized the opportunity to carry out high-end reconstruction programs, which greatly increased the price of real estate and encouraged an economic dynamism fueled by the return of wealthier populations (Peinturier 2020). This increased territorial fragmentation after a disaster and the diversity of territorial trajectories at the local level fuels the polysemy of a resilience that depends largely on the actor who envisions it. The majority of the white population of New Orleans declares that the crisis has been overcome, while the majority of the black population considers that the effects of the crisis continue to affect them strongly years after the disaster.

Thus, in addition to successive, closely spaced crises that feed territorial inertia by preventing effective (re)construction in the poorest countries (or on a local scale in better-endowed countries), crisis management actors can also (consciously or not) promote, through their actions, the long-term maintenance of deep vulnerabilities intrinsically linked to a country's political, economic and territorial system. More generally, in recent years, the emphasis placed on resilience by major actors (NGOs, donor organizations and states, and large companies) is more in keeping with a globalized neo-liberal system than it challenges it.

"Mobility", which is still a little-studied aspect of disasters, except for evacuation from the danger zone, is another dynamic component of crises, likely to export it and to complicate its spatial and temporal contours. Taking it into account to understand crises leads us to question the vocabulary used by actors and the victims themselves ("displaced", "evacuated", "deported", "transferred", "refugee" or "migrant"? Peinturier (2020)). This vocabulary shows how "disaster management plans very often limit the issue of mobilities only to the exit of victims from the danger zone" (Peinturier 2020), The issue of the long-term stay of "evacuees" and their return still remains difficult to consider as an integral part of crisis management for many actors (Fujiki 2017; Peinturier 2020) and as an element of restoration/reconstruction. Management and evacuation plans have been deeply improved since 2005 in New Orleans thanks to the feedback on the event and the dysfunctions observed in the management of shelters such as the superdome led to plan "to no longer leave anyone confined or refugee within the metropolis in case of mandatory evacuation", in particular by organizing the displacement of inhabitants without personal vehicle or particularly vulnerable (Peinturier 2020). Though the organization of the evacuation by bus was meticulously studied and organized, Peinturier, however, underlines certain inadequacies between the choice of location of the "17 recovery points, physically identified in the landscape and scattered over the territory, making it possible to serve temporary bus stations in the event of evacuation" and the geography of the neighborhoods, among the poorest in the city, in which "the number of households without a personal vehicle is higher than average, and which therefore constitute pockets of specific needs in terms of assisted evacuation".

Faced with crises characterized by sometimes blurred spatial and temporal boundaries, the stability or inertia of territorial frameworks (of data production to understand them or of management to deal with them) sometimes proves to be inadequate and can feed the crisis, amplify it or make it last, including by making it barely detectable. Thus, by examining the "blind spots in the data", Croisé (2020) shows how the invisibility of occupational cancers in Seine-Saint-Denis constitutes a complex, long-term social construction in which these cognitive barriers play an active role. The structuring of databases on industrial risks hinders the identification of links between the location of activities harmful to health and pathology. Biases in public archiving, transmission and access to information do not allow these links to be revealed or documented correctly. This invisibility is also fed by the progressive erasure of the material traces (in particular landscapes) of pathogenic places and activities (by destruction/reconstruction or through a patrimonialization of former industrial activities, including paradoxically within the framework of policies for a "sustainable and resilient" city). The disappearance or the change of function of places favors the oblivion of past and present risk (pollution) or makes the collective memory insufficient, ineffective to curb exposure (when these pollutions have overflowed from the initial site or when the site has not been cleaned up).

Croisé shows the distortions that exist between the reality of this risk of occupational cancers (based on the work and data of GISCOP 93 and 6 institutional databases) and representations that, by separating the environmental and occupational fields, have difficulty grasping the dynamic and systemic dimension of the new territorialities of these risks, which are still too often seen as static – circumscribed to the factory, the operating site – and not as dynamic via soil and water pollution, which shifts and modifies the geography of exposure to pathogenic products over time. The spillover of pollution outside the perimeter of the industrial site stricto sensu remains a blind spot in management, since “clean-up only concerns the company’s plots” (Croisé 2020). The very partial management of inherited pollution is highlighted through examples of renaturation or urban transformation (on former industrial sites in Aulnay-sous-Bois or Sevran) which show the paradoxical articulation between pollution and biodiversity within territorial branding-type development actions, and the way in which the “resilient city” approach has not taken on these problems which question its spatio-temporal framework (Croisé 2020).

7

Founding Crisis of Territorial Renewal

Yvette VEYRET[1] and Richard LAGANIER[2]
[1] Université Paris Ouest Nanterre, France
[2] Université de Paris, France

The introduction of the urban scale in the management of local and global crises is the result of a twofold observation that became apparent at the beginning of the 2000s: the world's urban population exceeded the rural population in 2007. At the beginning of the 21st century, 22 cities had a population of 5 to 10 million inhabitants, 402 had a population of 1 to 5 million and 433 had a population of 500,000 to 1 million. These different agglomerations are still growing rapidly, especially in developing and emerging countries. The concentration of people in cities, and, more particularly, in very large conurbations, raises two questions about crises and their management: (1) the concentration of industrial and tertiary activities, the energy requirements induced by transport, lifestyles and the functioning of the "urban metabolism" are major sources of a number of crises (climate, environmental, social, etc.) and imply that urban processes must be taken into account in policies to limit global crises, such as those associated with climate change. This involves reducing greenhouse gas emissions and defining "energy and environmental transition" scenarios; (2) cities suffer the consequences of crises because of the cumulative vulnerabilities inherent in urban dynamics (impacts related to extreme events, particularly in large urban areas and coastal cities; problems of securing water and food supplies for cities in arid and semi-arid developing countries; health impacts induced directly or indirectly by climate change, such as the effects of heat waves and the spread of epidemic diseases directly "stimulated" by climate change).

Territorial Crisis Management,
coordinated by Richard LAGANIER and Yyvette VEYRET.

Consequently, many current urban projects refer to the objectives of sustainable development and to the necessary adaptation of urban systems in the face of multiple local and global challenges relating to health, well-being, the environment, risks and crises. The goal is to think about the integration of hazards, from the neighborhood to the city, and of introducing complex thinking into the renewal of the urban project with regard to the multiple interrelationships between the economy, society and the environment that characterize them. This renewal is carried out by many actors, whatever the scale of the project: renovated housing, neighborhood renewal, design of new eco-neighborhoods or the creation of ecological cities. Among the challenges that are emerging is the definition of a post-carbon city that mobilizes transportation and housing as the primary sources of energy savings and seeks to mitigate the climate crisis and its impacts. It is also a question of defining a city that is resilient to risks in order to increase its capacity to deal with the multiple crises that affect it, and once again, habitat and urban technical networks are being questioned. In all cases, territorial transitions must be envisaged, whether they are submitted to and considered as a necessary response to one or more proven multiple crises or whether they are the subject of planning and decisions by anticipation to avoid possible future crises.

The strategies put in place to forge territorial renewal and bring about a certain number of changes are implemented in various ways according to different time frames. Risk management and crisis limitation require a reform in urban planning that can no longer dissociate networks from the territory. Land-use planning is in fact coupled with the practice of land management, defined by Michel Marié as "the capacity of technical and political institutions to self-regulate, that is, to constantly reevaluate the terms of their action according to the forces at play". This management is staged in reference to the notion of "associated environment", which insists on the interpenetration of technical objects and "technical and geographical environments". The product of such an operation consists of a sort of crossbreeding, a hybridization between techniques and societies (Dupuy 1991), and gives rise to different forms of adaptation according to different temporalities, as Jacques Theys (2009) invites us to do.

The short term is that of normative or incentive measures. It allows for gradual adjustments in urban housing, modes of transport, production methods and crisis management. The driving forces of change in territories are most often endogenous to the territory (an elected official promoting a project within the framework of their electoral mandate to limit the use of cars in the city and promote the use of public transport; an owner or a company that is sensitive to environmental issues and risks when renovating or building their house or place of business). They can also be accompanied by external incentives: national tax incentives aimed at reducing energy losses in public or private housing; regulations initiated by a public authority seeking to control the urbanization of municipalities facing major natural or industrial risks or the evolution of regulations at the national or even supranational level following natural or industrial crises (the factory explosion in Seveso, the La-Faute-sur-Mer disaster following the Xynthia storm and the associated coastal submersion).

On a multi-decade scale, the challenge is to transform the urban fabric and urban lifestyles. This challenge is based on major urban planning projects and investments and aims to offer a pleasant and safe city to citizens of the 21st century. Safety is indeed a combined whole, economic, social, health and environmental. The city has become the dominant social form because it is a place that favors protection as well as imagination, human creativity and easy economic and social exchanges. As such, it offers a more abundant supply of health services, both in quality and quantity, easier access to information and a larger and more diversified employment pool. The danger of implosion presented by unsustainable cities is, however, a global risk. The strong urban growth of the post-war period has consumed a lot of space, both for housing and for infrastructure, making the development of costly, impractical and inconvenient networks (roads, sanitation, electricity) unavoidable. Atmospheric pollution is thus a revelation of the qualitative and sanitary inconveniences induced by the city of the past, which is not very economical nor does it respect the health and well-being of its inhabitants. The increase in catastrophic events of industrial or natural origin also reveals the insufficient integration of risk into the city. The city is also a great complexity of communication, information and transport networks. This complexity creates a specific vulnerability. For some, the city is synonymous with security and enrichment, while for others, it evokes overcrowding, promiscuity, aggression, violence, epidemics, marginality, exclusion and pollution. These inequalities, both social and environmental, can only generate social tensions that may threaten the urban pact. We can therefore speak of a real urban crisis. This crisis can be identified, in part, with an image crisis, with a real urban identity crisis in a context of growing inequalities and territorial marking. Recovering a quality environment is a challenge, certainly in terms of image, but especially in that it is linked to territorial appropriation. Consequently, there are still major difficulties in renewing urban functioning in depth, starting with learning how to manage over a multi-decade horizon that goes beyond the duration of electoral mandates. In this perspective, major urban planning projects must think of space in terms other than those of jobs and economic profitability. Landscape and heritage management can be assets in their own right in the development of these projects. The transversality in the action of technical services will also have to go beyond outdated corporatism and will make it possible to better consider the complexity of the urban system. All this implies the necessary mobilization of the private sector as a corporate citizen.

Mobilized by this new urban governance, urban areas will have an essential role to play in dealing with multiple crises. The local councillors who were the guarantors of unchanging and precautionary management have been succeeded by a generation of elected builders who are seeking to reduce energy consumption by a factor of three or four, to significantly modify transport patterns and habits and to greatly reduce urban vulnerability to risks. It is to be hoped that tomorrow's "sustainable" cities will be those that are able to rediscover the democracy of the polis and well-being for all through a dynamic of shared projects.

In a more distant future, which some people relegate to the realm of utopia, the challenge is that of recomposing urban forms and functions. It implies major technological ruptures that are likely to have a significant impact in terms of carbon footprint or ecological footprint. These long-term transitions, which seek, for example, to move to a "decarbonized" energy model, are, in fact, much slower because they are subject to controversy. They are sometimes instrumentalized by supporters of the status quo and imply deep technological transformations that make a rapid transition impossible.

Consequently, renewing territories by integrating the lessons learned from past crises and risks means taking into account these different temporalities of action and designing urban projects that consider the complexity of territorial systems and their dynamics. This implies a global, multidimensional vision, based on the protection, prevention and adaptation of housing, neighborhoods, cities and urban infrastructures (networks, critical infrastructures, etc.) to hazards. It may also be a question of imagining forms of strategic retreat in the event that the dangers being combated lead to the definition of a resilience strategy that is based on the avoidance of extreme risk. The following examples seek to illustrate some concrete aspects of these strategic approaches, which envisage renewing the classical approaches to risk management.

7.1. Managing a risk territory on a global scale

The example of the Rhône basin shows the need to consider risks and crises in a global approach of development and management of territories. The management of the flood risk linked to flooding of the Rhône has moved from a protection that is sometimes insufficient, linked to dikes, to a global approach, an exemplary evolution of the integration of risk into planning and sustainable development practices. The floods of 1993–1994 and then of 2003 strongly affected the downstream Rhône valley and the Camargue in particular, these major crises are a sort of "founding act" of a new way of managing risk and trying to reduce the impact of new crises. Although they are debated since their presence can increase the vulnerability of the inhabitants and the facilities they are supposed to protect, dikes often remain indispensable, as evidenced by the river and coastal dikes of Belgium or the Netherlands (Peeters et al. 2013). This is the case of the Rhône dikes, which must contribute to reduce the river's divagations by raising its bed. Surrounded by alluvial rims, during large floods the Rhône can break these rims and flood vast areas. The dikes traditionally built and maintained since the Middle Ages proved to be more or less effective in the 19th century during the very strong floods of 1840 and 1856 (respective flows in Tarascon of 13,000 m^3/s and 12,500 m^3/s) when many breaches opened in the levees. After these events, the dikes were raised 1.50 m above the line reached by the 1856 flood. Changes in the bed of the Rhône in the 20th century, linked to the development of hydroelectricity, the development of navigation and the extension of irrigation (*Compagnie Nationale du Rhône*) have had consequences on the flow of the river. In addition, the decrease in agricultural pressure and the decline in agriculture during the 20th century, which was not marked by very high floods, explain why the dikes ceased to be monitored and maintained. It was only at the end of the century during the significant flood of 1993–1994 (9,500 m^3/s in Tarascon) that the situation became problematic once more.

7.1.1. *The age of risk management*

In the past, each owner built their own levee to protect their own property from flooding, nevertheless, "in 1805 (Imperial Decree of 4 prairial year XIII), the State agreed to contribute a quarter of the maintenance expenses of protection works, while placing the management of the works under the authority of the *Ponts et Chaussées*" (Bethemont 1972). The decree of 1849 allows for the creation of a unique management association for a given area, replacing the numerous existing associations. Following the catastrophic flood of 1856 and because of the extent of the damage, the state financed the repairs of dikes, which then made it possible to protect agricultural land. An imperial decree of 1856 recognized dikes as a "work of public utility". The maintenance of renovated dikes or their reconstruction was the responsibility of the owners of the land, the new dikes being managed by the "*syndicat des chaussées de grande Camargue*". At the end of the 20th century, the lack of maintenance was the rule, although small works were carried out with reduced financial means. The extent of the floods of 1993–1994, and then of 2003, justified a significant commitment from public authorities in securing the dikes as well as a different conception of protection and crisis management. In 2003, on the right bank of the Rhône, at the entrance to the river delta, the Beaucaire-Fourques plain suffered significant damage due to dike breaches. The northern districts of the city of Arles on the left bank were also flooded.

7.1.2. *SYMADREM*

Public authorities mobilized largely under the pressure of victims grouped in associations. Thus, in 2004, the *Syndicat mixte d'aménagement des digues du Rhône et de la Mer* (SYMADREM) was created, bringing together two regions (PACA, Languedoc-Roussillon), two departments (Bouches-du-Rhône and Gard), four municipalities in the Bouches du Rhône and eleven in the Gard. This syndicate replaces the *Syndicat intercommunal de gestion des digues du Rhône et de la mer* (SYDRHEMER) created after the 1993–1994 floods, whose area of application concerned only the left bank of the Rhône. The implementation of the "Rhône Plan" has strongly contributed to the structuring of SYMADREM, an organization equipped with the means and skills to fully assume the role of project manager on behalf of the communities. The functions of SYMADREM are related to the surveillance and maintenance of dikes, and it is responsible for the execution of works, structures and installations intended to ensure the protection of people and property against flooding from the Rhône and the sea. This syndicate now covers the entire lower Rhône river, including the delta, which is a particularly important aspect that allows us to consider the relationships between upstream and downstream, between the river banks and the "branches" of the delta. SYMADREM is responsible for major projects required by the state and local authorities, as well as measures of the "Rhône Plan" financed by the State-Regional Interregional Plan Contract (*Contrat de Plan Interrégional État-Régions*, CPIER).

7.1.3. *The Rhône Plan, a global vision*

SYMADREM is part of the "Rhône Plan" established after the floods of 2003. The many challenges of the territories crossed by the Rhône led the actors concerned to envisage a "federating tool" to establish a sustainable development project (see planrhone.fr). The "Rhône Plan" should contribute to "making the best use of the basin's potential, managing the constraints to which it is subject and preserving its environmental wealth and its inhabitants' quality of life". This strategy was developed in 2004 within the framework of a partnership between the state, the Rhône-Mediterranean Basin Committee and the three Regional Councils of Provence-Alpes-Côte-d'Azur, Rhône-Alpes and Languedoc-Roussillon, which were joined, in 2006, by the Regional Councils of Burgundy and Franche-Comté when the plan was extended to the Saône, as well as the CNR (*Compagnie Nationale du Rhône*) and EDF in 2014. The "Rhône Plan" therefore implies a global vision of the entire river, both in physical terms (hydrology, morphology, geology, etc.) and in terms of activities that must coexist as harmoniously as possible, as well as in terms of water quality, biodiversity and flood and crisis risk management. Thus, the

2005/2025 strategy insists on seven objectives: Rhône culture, floods, water quality, resources and biodiversity, energy, transport and tourism. With regard to risks, numerous operations have been carried out, in particular, work aimed at reducing flooding: reinforcement of dikes, reduction of vulnerability to flooding and the development of three flood risk management plans (upstream, middle and downstream Rhône).

The preservation of flood expansion fields and sediment management were programmed by the CPIER for the 2007–2013 period, as was the securing of dikes, which nevertheless came up against numerous obstacles: low mobilization of project owners, administrative difficulties in carrying out dike reinforcement projects (regulations relating to the safety of structures, the environment, etc.), the situation of orphan dikes ans financial problems. Reducing vulnerability was another objective for the 2007–2013 period, particularly regarding public buildings and networks. Living better with risk was the third component, which focused on risk culture and research-action. In order to prepare the 2015/2020 program, a strategic territorial diagnosis was carried out by an independent consulting firm. The partners of the "Rhône Plan" insisted on the central issue of flooding in relation to global warming and the implementation of the European "Flooding" Directive, which provides for "reducing the negative consequences for human health, cultural heritage and economic activity associated with flooding". The Directive requires a review of the risk prevention policy, since its objective is to identify priority territories for intervention where the highest stakes are concentrated (territories at significant risk of flooding, *territoires à risque important d'inondation* – TRI). Six TRIs of national importance have been recognized on the Rhône: Lyon, Vienne, Valence, Montélimar, Avignon and Arles, all of which are urban areas. The "Flooding" component for 2015–2020 insists on the need to act on the hazard; to reduce vulnerability; to live better with the risk and to focus on project management. The first objective requires the consideration of hydraulic structures (diagnosis, maintenance, security work). When the dike systems are fragile, as is the case locally downstream on the Rhône, it is necessary to consider reinforcing them. On the lower Rhône, the creation of a dike between Tarascon and Arles is envisaged, as is the securing of the dikes of the Grand Rhône (Salin-de-Giraud and Port-Saint-Louis-du-Rhône). The preservation of flood expansion fields justifies the proper maintenance of hydraulic structures allowing the functioning of flood plains (impoundment and drainage). Considerable work launched by SYMADREM aims to secure the cities of Beaucaire, Tarascon, Arles and Fourques in the downstream basin. However, such practices provoke conflicts emanating, in particular, from farmers who are not used to such practices since, in the 19th and 20th centuries, the priority of flood management concerned rural areas. Some farmers nevertheless accept the protection of urbanized areas at their expense, considering that they will have time to prepare themselves, thanks to a controlled arrival of water, and that they will also benefit from compensation (tax advantages and insurance coverage). Opponents of the project denounce the idea of discharges in the plains upstream of Arles and not in the Camargue further downstream, such a choice being considered as a way of protecting the political, economic and ecological stakes of the Camargue to the detriment of the protection of the goods and inhabitants of the upstream plains. Some dikes have been moved, thus allowing, for example, the "dredging" of the Petit Rhône. The raising of the industrial-port sites of Beaucaire and Tarascon, the treatment of the sensitive points identified during the studies of dangers and the securing of the Management Plan of Structures in Period of Floods (*Plan de Gestion des Ouvrages en Période de Crues*, PGOPC) were also carried out. In addition, the inhabitants have been mobilized in communal organizations in order to monitor dikes during floods and to remind people of the extent of past floods through awareness-raising actions. The surveillance and maintenance of dikes are the responsibility of the dike guards and the protection reserve, organized on a voluntary basis. This was legally recognized, within the framework of the law of modernization of civil security of August 13, 2004. Its operation and actions have been specified in relation to the SYMADREM's Flood Management Plan (PGOPC): tree cutting, manual clearing of brush in the vicinity of structures, motorized weeding of masonry, maintenance of accessories: cofferdams, etc. Teams of field agents, dike guards or volunteers from the communal civil security reserves of the communes along the Rhône, members of SYMADREM, regularly monitor the equipment in order to detect any weaknesses. SYMADREM manages 230 km of dikes thanks to the Spatial Information System on Dikes (SIRS, *Système d'informations à références spatiales sur les digues*) initiated by the National Institute of Research in Science and Technology for the Environment and Agriculture (IRSTEA, *Institut national de recherche en sciences et technologies pour l'environnement et l'agriculture*). This complex system makes it possible to list all the hydraulic structures existing in the Syndicate's area, which facilitates their monitoring, management and mapping (Mallet et al. in Royet and Bonelli 2013).

7.1.4. *Crisis management*

Crisis management on the lower Rhône depends on the alert level: above a certain threshold, a linear and continuous surveillance of the structures is carried out by the dike guards and volunteers from the communal civil security reserves. These teams, directed from the command post installed at SYMADREM, must prevent the appearance of breaches by monitoring the dikes. Equipping dikes with optical fibers and developing augmented reality devices to visualize flood-prone areas are the first links in an evolution towards "smart prevention" (Langumier et al. 2015). The monitoring of dikes by residents and flood markers could then give way to digital vigilance. In addition to the examination, maintenance and monitoring of the condition of dikes, the PGOPC must improve crisis management and, particularly, emergency interventions during floods. The management plan takes into account the evolution of the flows of the river and its tributaries. A "Flow Correspondent" post has been created whose manager informs the command post of the river flow forecasts announced by the Flood Forecasting Service[1] (*Service de Prévision des Crues*). In order to test the implementation of the PGOPC "in real life" and in accordance with the regulations, simulation exercises are carried out regularly. The flood of February 2009, which justified the triggering of alert 2 (on three levels) and the daytime monitoring of the structures, also allowed the general organization of the PGOPC to be tested. The COGMP includes signage for monitoring and emergency response teams to facilitate access to areas to be monitored when repairs to degraded access roads are required and new roads need to be built. A system for identifying malfunctions has been developed. based on more than 800 markers. installed every 250 m or so on the dikes. Storage areas for materials that are necessary for emergency interventions are also located along the Rhône dikes (Bravard et al. 2008).

7.1.5. *Evolution of the legislation, the future of SYMADREM*

The transfer of the management of aquatic environments and flood prevention (GEMAPI, *gestion des milieux aquatiques et prévention des inondations*) competence to EPCIs (public establishments for intermunicipal cooperation), as of 2018, could lead to an evolution of the competent syndicate in two regions and two departments (Gard and Bouches-du-Rhône), or even the questioning of SYMADREM. Will these regions and departments be able to continue to finance SYMADREM while the NOTRe law assigns financing to the communes? Article 59 of the law of January 27, 2014, on the modernization of territorial public action and the affirmation of metropolises allows the regions and departments participating in mixed syndicates competent in GEMAPI to continue their commitments until January 1, 2020 at the latest. Beyond that date, the intervention of regions and departments will remain possible, but only in areas that do not fall within the exclusive scope of the GEMAPI competence. Regions and departments will therefore be able to continue to intervene as members of mixed syndicates competent in GEMAPI in order to undertake the study, execution or operation of hydraulic works that are essential to civil security (emergency drinking water supply, firefighting), as well as the operation, maintenance and development of existing hydraulic works when the latter have functions other than flood control.

1. http://www.symadrem.fr/wp-content/uploads/2015/05/pgopc-1.pdf.

Finally, the general code of local authorities allows the departments to continue their contribution to the financing of projects for which the municipalities or their groupings are responsible, at their request. Therefore, public establishments of intercommunal cooperation competent in GEMAPI will be able, if they so request, to continue to benefit from financial support from the departments for flood control beyond January 1, 2020. Nevertheless, as far as SYMADREM is concerned, the Provence, Alpes, Côte d'Azur and Occitanie regions as well as the department of Gard have withdrawn, while the department of Bouches du Rhône has remained in the syndicate. However, agreements have been established so that the work programs included in the CPIER 2015–2020 and in the program to secure the works continue to be financed, in part, by the regions (but the allocation of the Occitanie region has been reduced from 40% to 30%) and the department of Gard. The intermunicipalities replace the communes within the structure. Thus, the management of floods on the Lower Rhône and in the Camargue shows significant developments, the work on the scale of this vast geographical area, included more broadly in the "Rhône Plan" has initiated an effective way to consider its management, the existence of a single decision-maker representing regions, departments, municipalities were to avoid that each entity does not decide, in its own way and independently, how to manage risks and crises. The question of governance is therefore central.

Although the "Rhône Plan" is an indispensable example of global management of a territory, in many cases, risks or crises are still taken into account through specific and sectoral responses, which are certainly necessary but which should systematically be included in a broader vision.

7.2. Reducing risk through protection

To reduce the risks and scale of crises, the responses are often still based on protection by means of adapted developments. In the case of floods, the implementation of dikes is old, supplemented by dams, polders and so on. On the coast, reducing danger means locally leaving room for hazards, in particular, by "depoldering". Using solutions that are based on the functioning of nature and ecosystem services is the solution now systematically put forward. From now on, hydraulic and ecological functions are no longer perceived as antagonistic, but complementary in the perspective of sustainable management of watercourses and flood control (Bravard and Clemens 2008), as shown by the equipment of river polders.

7.2.1. *Dikes and diking systems, regulation, management*

Protection facilities, especially dikes, were, historically speaking, the first responses to floods. The legislation concerning protection and dikes has evolved a lot. The law of January 27, 2014 proposes, for the first time, a regulatory definition of dikes envisaged as "works built or developed to prevent flooding and submersion". This law distinguishes dikes from hydraulic developments, which have a different function and generally meet several uses. Dikes and hydraulic installations are also sometimes a source of aggravated dangers. The security created by dikes leads to the densification of buildings and activities behind these protections. If the dikes fail or are overtopped, the associated flooding is much more dangerous than the flood from which it was intended to protect the population.

River and maritime protection dikes began to interest those involved in planning and risk management in the 1990s, within the framework of two circulars, one in 1994 and the other in 1999, which called for a national inventory of structures and their managers. Although river dikes were integrated into the 2003 Water Act (authorization or declaration system), this was not the case for maritime dikes, which thus escaped the first dike safety control, which established the notion of "classification as being of interest to public safety". Dikes protecting against marine flooding were not taken into account until 2007, when a classification of dikes (A, B, C, D) was established, according to the population to be protected and the height of the dike. The treatment of dikes is now based on that of dams, a hazard study must be conducted by those responsible for the works.

The multiple regulations concerning dikes have been specified within the framework of the GEMAPI law of 2014 and in particular paragraph 5 of the law "defense against floods and against the sea", which concerns protection works and dikes in particular but does not take into account crisis management. However, on the scale of France, the dike system is still imperfectly known. The total length of dikes listed in the state's database (*Système d'information sur les ouvrages hydrauliques*) amounts to approximately 9,000 kilometers in metropolitan France. These dikes may be managed by private individuals (individuals, companies) or public entities (municipalities, public establishments for intermunicipal cooperation, the state and its institutions, departments, regions, authorized syndicate associations, etc.), although some do not have managers. In 2014, for class B dikes (in other words, according to the 2007 decree then in force, those protecting a population of more than 1,000 people and exceeding 1 m in height), 24% of these dikes were managed by local authorities (municipalities or EPCI-FP), 27% by mixed or intermunicipal syndicates, 16% by companies, 13% by the state, 7% by ASAs (*Associations syndicale autorisée*, Authorized Syndical Associations), 1% by private individuals and 12% remained without an identified manager. The case of "orphan" dikes, that is, dikes whose owner is not known, is problematic. It is therefore essential to find out who owns the land bearing the structure, and one of the objectives of the GEMAPI competence is precisely to find managers for structures that are currently without an identified manager and are in poor condition. The Civil Code specifies, with regard to escheated estates and property without an owner, that ownership reverts by low to the municipality or, if it refuses, to the public establishment of intermunicipal cooperation (EPCI, *établissement public de coopération intercommunale*) and, in the event of renunciation by the latter, either to the *Conservatoire du Littoral* or to the state. The decree of May 12, 2015 specifies the terms and conditions for managing dikes. From now on, EPCI-FP are obliged to manage dikes and can delegate or transfer this competence to syndicates (EPAGE: local authorities constituted as a mixed syndicate on the scale of a coastal river catchment affected by recurrent flooding or of a sub-catchment area of a major river) in order to ensure the prevention of flooding and marine submersion as well as the management of non-domestic watercourses. This decree insists on the notion of a diking system composed of one or several dikes that are intended to defend an area against floods and/or submersions as well as any other work contributing to protection: gates, flap gates, road/SNCF embankments, walls, that is, works which initially were not intended to protect against floods but which can ensure a protective function. All diking systems or hydraulic developments involve a regularization or (environmental) authorization procedure, whether they are new or already existing. In application of the 2015 decree, simplified authorizations have been granted (without public consultation) for already classified works. To install, consider the effectiveness of and manage a diking system that is intended to defend a territory against floods and submersions, the regulation requires the completion of a hazard study (2015 decree) which concerns the diking system and not, as provided for in the previous regulation, only dikes. The purpose of this study is "to show which area is protected by which system and for which hazard, which watercourse or which marine submersion and up to which height". The hazard study is a risk analysis taking into account both the environment and the characteristics of the works of the diking system, their behavior and their management methods. The approach to be followed in order to carry out the hazard study is specified by the decree, it must show the interest and the efficiency of the hydraulic works selected to ensure a defined level of protection on a given territory. Four scenarios of water rise, according to the magnitude of the events responsible (flood, storm), are considered: the first in normal operation, the second in the case of functional failure (absence of cofferdams, for example), the third in the case of structural failure (breach) and the last in the situation of a hazard reaching the maximum level foreseen in the risk prevention plan (PPR, *plan de prévention des risques*). Graphic and cartographic representations accompany the hazard study. This study and its update must be carried out by an approved organization (for a fixed period), defined by the order of February 15, 2018. A local authority can be accredited as shown by the example of the General Council of Charente-Maritime; it is also the case of a syndicate such as the mixed syndicate of development of the Durance valley. The aim of these new approaches is to identify a single project owner or manager of the diking system. In fine, they assume responsibility for the content of the hazard study with the state services and its control services. They thus have a major interest in the proper conduction of the study. In this context, the legal and financial responsibility of the managing authority, that is, the GEMAPI authority, is not engaged. Thus, if the hazard study has foreseen that the structure could provide protection against a 10-year flood, the manager cannot be held responsible for damage caused by a 100-year[2] flood.

7.2.2. The equipment of rivers with polders

Polders along rivers represent a radical change in water and flood risk management. It is a matter of retaining water when the river flow increases, storing it and then releasing it slowly when the flow decreases. These practices are compatible with the maintenance or reconstitution of the biodiversity of the banks and the major bed. Along rivers, two types of polders can exist. Summer polders, separated from the minor riverbed by a low dike called a "summer dike", are put in the water by submerging this dike. Generally, there is no water intake or emptying structure. The submersion takes place during low-amplitude floods. The retention polders have maneuverable water intake and discharge structures. The opening of the water intake when the water level rises provides effective protection against flooding further downstream. The Rhine polders, which are intended to reduce flooding downstream of their location, require an agreement between France and Germany. It is therefore necessary to consider the watercourse not only in the municipality or the community of municipalities it crosses, but in the whole catchment area. Financed by the Franco-German Convention of December 6, 1982, the creation of the Moder polder is part of a large Franco-German program of about twenty developments; like the Erstein polder, all of them must contribute to regulating the Rhine floods. The system also responds to the need to preserve and even restore the alluvial forest. However, the correction of the Rhine bed in the 19th century (1840–1876), which consisted of shortening and fixing the minor bed between submersible dikes and delimiting a new major bed by unsinkable dikes spaced 1 to 2 km apart, profoundly modified the river dynamics. With the regulation carried out during the first half of the 20th century (1906–1960), flooded groins were implanted in the minor bed, which was cluttered with numerous sandy and

2. https://www.cepri.net/tl_files/Guides%20CEPRI/Guide_gemapi_PI.pdf.

gravel banks. A channel deep enough for navigation was installed. Finally, between 1928 and 1977, 10 hydroelectric plants were built on canalized reaches or on the corrected bed. These works caused a strong incision upstream of Neuf-Brisach, the disappearance of the braiding and the disconnection of the channels and associated islands, these anastomoses called "Giessen" on the Rhine, the suppression of floodable zones, the acceleration of flood waves, the rise of flood peaks downstream of the canalized sector (International Commission of the Hydrology of the Rhine Basin 1977; Carbiener 1983; Dister 1992; Schmitt 1995, cited by Boudjellal et al. 2013). In other words, the developments were the cause of hydrological, ecological and economic problems, particularly during floods, which in 2004 caused damage estimated at 6 billion euros on the Upper Rhine. Faced with this situation, France and Germany undertook to restore, downstream of the developed sectors of the Rhine, the level of protection existing before the canalization (Franco-German Convention of 1982). To this end, each of the two states has decided to build retention areas along the Rhine with a total volume of 270 million m^3. The objective is to stop major floods so that the flow does not exceed 5,000 m^3/s at Maxau (Lauterbourg sector) and 6,000 m^3/s at Worms.

The treatment of the river to reduce the important – bicentennial – floods and to allow the ecological restoration of the river and its banks requires an agreement between the actors of the two banks. Different types of measures have been implemented in France: stopping the turbines of the EDF plants between Basel and Strasbourg so that the water of the canalized Rhine flows towards the old bed; the realization of two polders with a total capacity of 13.2 million m^3: the Moder polder (operational since 1992) north of Strasbourg, and the Erstein polder (completed in 2003), south of Strasbourg. In Germany, 16 polders should hold 140 million m^3. The French polders are managed by *Voies Navigables de France* (VNF), but Franco-German operating instructions for the Rhine structures are gathered in what is called the "*Classeur Gris*". The "Rhine Plan" sets out the specific ORSEC provisions. It relies on several types of actors, including the prefect who decides on the triggering of the most important measures and who is advised by the DREAL and the DDT, the operators who inform and implement (mainly VNF and EDF) the recommended actions.

The Erstein polder, which is located 15 km south of Strasbourg, is located in the sector of the Rhine known as in braids and anastomoses or "Giessen" (Neuf-Brisach – Strasbourg), with a slope of 0.5–0.8‰ (Carbiener 1983), it occupies 600 ha, largely wooded, and has a storage capacity of 7.8 million m^3. Located on the left bank of the Rhine, this polder is a pilot site designed to reconcile flood control, the safety of people and property downstream of the canalized Rhine and protection/restoration of alluvial ecosystems. The Erstein polder is protected by a system of dikes that were set up during the construction of the Strasbourg waterfall in 1970. These dikes were built to completely encircle the polder. The polder is impounded by two intake structures located in the dike of the canalized Rhine, their maximum capacities being 220 m/s^3 (main intake) and 30 m/s^3 (secondary intake). Energy dissipation at the outlet of the structures is ensured by specially dug basins designed to avoid erosion problems. Depending on the size of the flood peak, the filling lasts 13 to 15 hours. The total duration of submersion, including filling and emptying, should not exceed 20 consecutive days (emptying takes place in two to four days). The operation of the polder allows the "Giessen" to regain a hydrological and ecological mode of operation that is quite similar to that which existed before the Rhine was canalized (Carbiener op. cit.), namely, a phreatic supply in periods of low water and a mixed supply combining river water in periods of high water. In addition, once a year, the polder experiences an "ecological flooding" that allows the restoration of alluvial ecosystems. In 2007, from August 9 to 10, the Erstein polder experienced its first real flood retention when the flow of the Rhine reached 4,147 m/s^3. The German and Moder polders were not called upon. However, the Erstein polder represents only 3% of the flood control capacity necessary for a 200-year protection of the populations between Iffezheim and Worms[3].

In estuaries, old polders can become discharge polders in the case of high tide and/or storms, in Belgium they constitute "controlled reduced tide flood regulation zones". In these polders, the dikes are equipped with sluices, to allow entry by overflow of the estuary waters in the case of a storm or exceptional tide, the exit of the waters at the end of the event is controlled (Peeters et al. 2013). This retention of water reduces the height of the tide upstream of the polder, so, on the Scheldt between Antwerp and Ghent, 580 ha of polder are used in order to divide the extent of flooding by five.

7.2.3. *Soft management of coastal risk areas*

The soft methods of coastal management lead to evoking forms of "coastal resilience" defined as "the capacity of coastal systems to adapt to disturbances". These methods (also called soft methods), avoid the fixation of the coastline as much as possible while promoting its growth (Lenôtre 2009).

7.2.3.1. *Beach nourishment and beach drainage*

Beach nourishment involves a source of sand that is close enough to help dissipate wave energy and marine flooding, accompany sea level rise and protect

3. http://www.set-revue.fr/sites/default/files/articles-eat/pdf/DG2008-PUB00026460.pdfi.

seawalls. However, this practice comes at a cost; it involves finding sand without disturbing the area from which it is taken. Another aspect of soft protection is the restoration of dunes (thanks, in particular, to ecological engineering) and their fixation by plantations and ganivelles. The drainage of beaches is a more original practice; this method appeared in the United States in 1940 and consists of draining the water table under the beach to compensate for the annual recharging of the beach. This technique, called "ecoplage", mobilizes one or more drains installed parallel to the coastline, through which the wave water infiltrates, thus ensuring a desaturation of the sand and allowing a good capture of the sediments on the top of the beach; collecting wells and transport pipes allow the water captured by the drains to flow, by gravity, to a pumping station. This station evacuates the filtered seawater, which can be either discharged into the sea or used for nearby activities (desalination unit, swimming pool, aquarium, thalasso, fish farming, etc.). An assessment carried out 10 years after the implementation of an ecobeach in Quend, on the Picardy coast, shows its effectiveness in draining the beach. More than 20,000 m^3 of sand were recovered from the beach at the foot of the dike (Lambert et al. 2007).

7.2.3.2. *Depoldering*

Depoldering means "bringing the sea back into a land that was once diked", that is, transforming a polder to revive a maritime marsh made up of mudflats and salt meadows. It also means modifying and moving back the coastline, which is usually considered as fixed by many coastal actors. However, the primary characteristic of the coast is its mobility at all scales of time and space. The fixist conception of the coastline constitutes a brake, in particular, for the disappearance of certain polders, spaces which are nevertheless attributable to anthropic actions but which are sometimes perceived by local residents or other actors as intangible. A contradiction then appears between the recent desire to recover "natural" seafronts locally and the image of spaces abandoned to the forces of nature, a place of unacceptable wastelands following cultivation (Bazin and Olivry 2017). Thus, in Brittany, many small marshes, dammed in the 19th and 20th centuries for productive purposes, are now in disuse, because agriculture has been declining since the 1950s in the polders of Brittany's Finistère. The same is true on the banks of the Ria Formosa, in southern Portugal, where the marshes, dammed up in the previous century and intensively cultivated until the 1960s, are now practically all fallow (Baron-Yellès et al. 2002). The reasons for this abandonment are mainly related to the general reduction in demand for agricultural land, as well as to the constraints of use of these areas that are reclaimed from the sea and always threatened by it. In addition, from the 1980s onwards, diking was generally contested in the name of ecology, particularly in Germany, where nature conservationists highlighted the destruction of mudflats and salt meadows with their rich biodiversity in order to create polders.

Consequently, some polders are voluntarily returned to the sea, within the framework of a new policy of "depoldering", notably in Great Britain. Depoldering can be done by voluntarily opening gaps in the dikes so that sea water can enter the polder. This is the case in Noirmoutier in the Sebastopol polder, where the entry of sea water remains regulated in its extent or duration. Some dikes are also voluntarily cut down to obtain the immersion of the polder; this is the case of the Aber de Crozon in France and Brandcaster in Great Britain. Sometimes breaches can also appear in the dikes because they are no longer maintained. Lydie Goeldner Gianella notes, however, that the depoldered surfaces are quite small, 90 km^2 in 2014, and close to 180 km^2 in 2018 (Goeldner-Gianella 2018).

The reasons for depoldering are multiple; it can be to restore the ecosystem of maritime marshes to the state before reclamation and the associated biodiversity, but the strategy also resonates with the challenges related to climate change and its impacts, including those in terms of sea level rise (Goeldner-Gianellla 2012). Thus, one of the main objectives of depredation in Great Britain is the protection of nature and the restoration of biodiversity. Nature protection associations (Royal Society for the Protection of Birds, etc.) play an important role in depoldering, especially since they may own the dikes or even the polders themselves. Such possibilities are not available in other European countries, where the foreshore is sometimes publicly owned (Goeldner-Gianella 2007). Depoldering can also help to reduce the consequences of rising sea levels, which itself is a consequence of global warming. This so-called defensive depoldering then becomes a possible form of risk anagement. It exists especially on the coasts of the North Sea, Great Britain, the Netherlands and Belgium, where violent storms are frequent and dangerous. Great Britain, through "managed realignment", is a pioneer in this field. Depoldering is becoming a "soft" means to adapt to the rise of the sea level. The waves that reach the salt meadow are slowed down, their energy is reduced and the height of water that reaches the dike is lower. This reduction in wave strength is an element of integrated coastal zone management. Thus, depoldering can combine defense against risks and ecological restoration, which is in line with the objectives envisaged for the implementation of storage polders along rivers. Depoldering sometimes has other justifications; it can result from the implementation of a compensatory policy leading to the recreation of maritime marshes to compensate for the same surface area destroyed elsewhere, particularly during the development of a port. In France, depoldering can be carried out by the *Conservatoire du Littoral*, one of the bases of its action is to leave coastal areas free, not to resist the sea in order to preserve the ecological characteristics and natural appearance of the sites. The strategy of the Conservatoire is part of the national strategy of integrated management of the coastline, in close collaboration with all the actors of the territories. The Conservatoire intervenes in sectors undergoing strong evolution in order to contribute to the consolidation of buffer zones between land and sea. Its choice of areas to be treated is based on several aspects, including risk management, ecosystem quality, landscape value and cost.

Depoldering must meet very specific objectives, so the necessary securing of the south of the Bay of the Somme to cope with the risk of flooding, as evidenced by the existence of the PPR of the Bas Champs, approved in 2017 and which concerns the municipalities of Cayeux-sur-mer, Lanchères and Pendé, can be a major reason for depoldering, accepted by local stakeholders. Added to this is the fight against silting of the port of Hourdel. The syndicate committee of the *Syndicat Mixte Baie de Somme-Grand Littoral Picard* (SMBSGLP) has decided in favor of depoldering the basin bottom. However, on the French coasts, depoldering remains the exception, especially since recent scientific research tends to reduce the role of depoldering in the face of flooding hazards. Indeed, the capacity of mudflats to attenuate swell decreases with the increase in water height during storms. Depoldering therefore seems to be a very limited brake on large storm waves (Niemeyer and Kaiser 2001). It is still necessary to maintain or even raise the dikes that protect the hinterland. In some cases, depoldering requires raising a dike or building one to isolate the submerged polder from the hinterland (Goeldner-Gianella 2007, 2018). Overall, depoldering remains quite limited. There is often a great deal of opposition from local residents, particularly for landscape reasons, while the ecological benefits of depoldering are not well known and are not sufficiently explained.

Numerous solutions exist to reduce the importance of hazards, and today they are almost systematically associated with other objectives, notably landscape ones, as well as being intended to protect or restore a rich biodiversity. Today, protection is largely part of regional planning policies that also take into account the way people live and their unequal vulnerability.

7.3. Adapting the habitat

Facing technological and natural risks requires an adaptation of the building. Building security must be adapted to various hazards, with responses varying according to the different types of flooding (slow or rapid torrential flooding, coastal flooding waves), earth movements, earthquakes and also technological risks (explosion, pollution, fire, etc.). Regarding GHG emissions and energy savings, several technical solutions have been put forward.

7.3.1. *Responses to natural hazards*

7.3.1.1. *Habitat and flooding*

The first response to the risk of flooding has always been to avoid settling in areas subject to flooding. In the past, many villages were built on hillocks or on terraces overlooking the major river beds, but, over time, urban development, land pressure, sometimes forgetting the danger, and the attractiveness of rivers or the seashore have led to construction in flood-prone areas. According to INSEE in 2016, in France, 1 million inhabitants live in potentially floodable areas, to which must be added more than 100,000 tourists staying in campsites or hotels that are exposed to the risk of flooding. The risk also concerns companies: 28% of employees work in establishments located in flood-prone areas. Many facilities in flood-prone areas predate the Flood Risk Prevention Plans introduced in 1995 (Barnier Law), so solutions must be found to reduce the risk (CEPRI 2010; Salagnac et al. 2014). Depending on the situation of the habitat and its characteristics, the safety of people in flood-prone areas requires either the practice of "dry floodproofing" which consists of preventing water from entering the building (installation of cofferdams, sandbags), although this solution is not always possible, or to consider "wet floodproofing" necessary when there is a strong pressure of water against the building; it is therefore necessary to balance the pressure between the exterior and interior, which implies foreseeing the arrival of water into the building (waterproofing all the networks, etc.). Recent regulations in France require that homes in flood-prone areas have a refuge platform, an emergency exit, especially through the roofs and always located above the highest known water level (PHEC, *plus hautes eaux connues*). This exit must be easily accessible by a staircase or an interior ladder, including at night, in the absence of lighting, must be associated with acceptable safety conditions (the ability to call or sign to the outside) with a minimum of comfort (sufficient space on the roof), and with the possibility of intervention of the emergency services and the evacuation of people. The existence of the PPRI (Flood Risk Prevention Plan) has not always prevented construction in floodable areas; in this case, it must be adapted to the situation and present a first floor that is located above the PHEC or above the level of the 100-year flood, if it is considered higher than the PHEC. This method of construction is valid for individual houses as well as for collective housing. The void thus left must remain so. Thus, in the Parc-en-Seine operation in Villeneuve-le-Roi, the living level is 14 cm above the highest known water level (PHEC), and the first floors are used for parking lots that are built in excavations and open to allow the circulation of water during floods[4]. Nevertheless, despite the pilings, if the networks (electricity, water,

4. https://www.cepri.net/tl_files/pdf/guidevulnerabilite.pdf.

sanitation) supplying houses and buildings in these flood-prone areas are not adapted, the inhabitants will not be able to stay in their homes during flooding.

7.3.1.2. *Seismic housing*

The modes of adaptation of buildings also necessary to face earthquakes are sometimes old; thus, the Cretan constructions of the Minoan period, which were made in light materials, established systematically on rigid substrates, at the bottoms of slopes and not at the tops of hills, also presented decroachments of frontages that allowed such buildings to resist in the event of earthquakes. Inside, the division of the space into small independent units delimited by corridors, interior courtyards and so on made each block autonomous during earthquakes (Poursoulis et al. 2000). More than 2,500 years ago, the Chinese used wooden joints that favored the flexibility of structures; these "dougongs" dissipated energy by friction. Similarly, wooden joints used in the Middle East for the construction of the Petra temples give certain flexibility to the structure, each floor being able to behave "independently", thus reducing constraints. Today, several simple techniques exist to construct earthquake-resistant buildings, including avoiding asymmetries that cause areas of weakness in the building. The materials used must be of good quality; this is the case of concrete and metal bars reinforcing the concrete slabs. Small buildings, and those of a single block, must have horizontal and vertical "chainings"; by linking the different sections of walls, we obtain a "box" that is able to support stress. Another technique consists of "anchoring" the building in the ground by using deep piles, which must reach a resistant substrate. Large buildings must be able to move without breaking, so buildings cannot be fixed to the ground, but built on jacks and have flexible joints. Seismic construction represents an additional cost of about 5 to 10% compared to the cost of a "normal" building, but the reinforcement of buildings that are already built is much more expensive and can only be done on small buildings for which it is necessary to add structures, "sills" – reinforced concrete beams that are arranged horizontally to distribute the loads of the supported elements (walls, columns, slabs) or/and to transfer them to point supports – it is necessary to consolidate areas of discontinuity. In Japan, where seismic practices are old, more and more complex systems are used to protect high-rise buildings in particular: jacks, springs, rails, water tanks, stays and rubber rings. The Japanese legislation on construction standards, revised in 2000, defines the criteria to "limit material damage and save a maximum of human lives". During earthquakes, buildings deform because the foundations move, and the higher they are, the greater the deformation. The choice of technologies depends on the height, mass and architecture of the buildings, as well as the cost. Rubber boots and oil dampers are frequently used.

The new seismic zoning of France entered into force in 2014 and revises the old zoning of 1991 that was established based on known earthquakes in the past. This new zoning responds to scientific advances and the new European earthquake-resistant construction code (Eurocode 8 or EC8). It is based on a probabilistic approach that consists of calculating the probability of occurrence of a given seismic movement in a given location and over a given time, knowing that Eurocode 8 recommends 475 years as the return period: “this probabilistic study is based on all known seismicity (from magnitude 3.5-4), the return period of seismicity, the seismotectonic zoning (i.e. a division into source zones where seismicity is considered homogeneous)” (Gaspar 2016, cited in Cartier and Valette 2016). The new regulation multiplies the number of municipalities affected by seismic requirements by four (approximately 21,000). Even synthesized, the whole document remains quite complex to approach, moreover, building professionals have challenged the multiplication of seismic zones and consequently the additional costs of construction. Thus, “[…] eight professional organizations of the construction sector have asked political authorities for simplifications and proposed a hundred […] adaptations or deletions of regulatory and normative requirements […]”. These organizations admit that “technical regulations and standards represent factors of quality improvement and progress in construction”, as much in terms of performance as in terms of safety or health, but they denounce “the increasing and sustained piling up of standards which participates in the deterioration of the economic situation of construction” (Gralepois 2012). Strong pressures have led to a reduction in the scope of the Eurocode, henceforth, the decree is limited to non-structural elements of the built environment that present a risk to the safety of persons because of their weight, their methods of fixation or their location; in addition, the seismic regulations must only be mobilized in the case of heavy work on the building[5]. The law of July 13, 2006 on the national commitment to housing has extended the possibility of controlling and sanctioning non-compliance with seismic construction rules. The owner must provide a certificate from a technical controller certifying the respect of these rules[6]. Many researchers and users denounce the complexity of the Eurocode; nevertheless, the Eurocode should contribute to greatly reducing the effects of a seismic crisis, although its scope is reduced by the pressures exerted in the name of economy and to the detriment of safety.

7.3.1.3. *Habitat and ground movements*

The diversity of ground movements requires adaptation to the precise type of movement. Thus, protection against the shrinkage and swelling of clay requires relatively simple measures to adapt buildings: deep foundations, reinforcement of the building structure. The reduction of water evaporation from the ground and the

5. BRGM. http://www.planseisme.fr/Batiments-a-Risque-Normal.html.

6. https://www.cohesion-territoires.gouv.fr/construction-et-risques-sismiques.

removal of vegetation (especially trees) near foundations are also recommended. Construction in zones that are sensitive to the collapse of underground cavities (fontis) is more complex: the lives of the occupants can be put at risk, as shown by the example of Clamart, where, in 1961, six hectares of chalk quarries collapsed to a height of two to four meters. Six streets disappeared, and the toll was 21 dead and 45 injured, with more than 273 victims and 23 buildings destroyed. The existence of cavities often corresponds to former quarries; this is the case under Paris, where they are monitored by the General Inspection of Quarries (IGC, *Inspection générale des carrières*). Cavities are numerous in other French regions (Normandy, Picardy, etc.) and in former mining regions (the north and east of France). As soon as an underground cavity is identified, its stability needs to be clarified, if a development is to be carried out on the surface or if this surface is already occupied. Underground risks, which can constitute an obstacle to urban development, induce various planning choices, excluding a unique answer and involving technical services as well as administrations, elected officials, insurers and the population (Veyret and Bouchon 2002). The municipalities or their groupings that are competent for urban planning documents draw up maps delimiting the sites that are located above underground cavities (Article L563-6-I of the environmental code); these areas that appear in the graphic regulations of the PLU therefore imply the prohibition of construction or the imposition of special conditions for construction or any other development. The PLU presentation report must then explain the zoning and the applicable rules. In the case of new constructions, protection measures are strict. They imply reducing the risk of settlement by anchoring the buildings deeply in a stable rock located at the base of the cavities or by treating the cavities themselves (direct filling, hydraulic filling, mortar or grout injections). The choice between the different techniques depends on the configuration of the site and the improvements that have been made or are to be made on the surface. In the most dangerous situations, the risk prevention plan (PPR) may lead to the expropriation of the inhabitants, as was the case in Chanteloup-les-Vignes on the Hautil gypsum massif. At the end of the 20th century, the cost of consolidating these houses was considered to be three times higher than the estimated value of the property concerned.

7.3.2. *Housing and technological risks*

The law of July 30, 2003 on the prevention of technological and natural risks, which introduced the Technological Risk Prevention Plans (PPRT, *Plan de Prévention des Risques Technologiques*), renews the tools for controlling urban planning. The PPRT imposes adaptation work on existing buildings and the implementation of appropriate construction standards for new buildings in the blue zones, "zones subject to restrictions corresponding to an average level of risk, requiring however restrictions adapted to the type of stake". In Jarrie, in Isère, two companies classified as SEVESO "high threshold", Arkema and Cezus, manufacture and process chemical products (hydrogen peroxide, bleach, etc. for Arkema, zirconium sponge, hafnium, magnesium and silicon salts and oxides for Cezus). These activities can be the source of a toxic cloud (chlorine), soil and water pollution (discharge into the Romanche River) and a BLEVE (boiling liquid expanding vapor explosion) linked to the storage of methyl chloride, a highly flammable liquefied gas. Approximately 800 private homes and a hundred or so homes belonging to social landlords are subject to the risks indicated. Responses in terms of housing must therefore be adapted to very different dangers. In the case of toxic effect, the houses must have a containment space, which means, in particular, that the air inlets and outlets must be sealed off. Against the overpressure effect, it is necessary to reinforce the woodwork while the thermal effect requires considering the structure of the house itself. At the date of approval of the PPRT, the owners of the plant are required to ensure the protection of the occupants and their property for the various types of hazards envisaged in the PPRT regulations. Thus, technological risk prevention work in existing housing requires information on the various hazards to which the building in question is exposed and knowledge of the objective to be achieved in terms of safety according to the type of risk. In the case of new constructions, the Center for Studies and Expertise on Risks, the Environment, Mobility and Development (CEREMA) has produced a guide for designers of new housing to help them take into account the requirements of the PPRT. The General Directorate for Risk Prevention (DGPR, *Direction Générale de la Prévention des Risques*) has set up experimental support programs for PPRTs, called Industrial Risk Support Programs (PARI, *Programmes d'Accompagnement des Risques Industriels*), designed to help local residents, local authorities and industrialists implement the work prescribed by PPRTs. One of the aims of the PARIs is to simplify the procedures required to carry out the reinforcement work prescribed in individual or collective buildings for residential use. The General Service for Scientific and Technical Research (DGRST, *Délégation générale à la recherche scientifique et technique*) is experimenting with a PARI at eight sites classified as Seveso: Jarrie, Marseille Saint-Menet, Lespinasse, Bassens, Roussillon, Donges, Lanester and Strasbourg. A one-stop shop is used to pay the various contributions. The financing of home security work is provided by a tax credit whose rate is 40% of the amount of work, with a ceiling of 20,000 euros or 10% of the market value of the property and, since the law of July 16, 2013, a contribution of industrialists and local authorities up to 25% each is added to the tax credit. Assistance to individuals can therefore reach 90% for the realization of the work of making their housing safe.

7.3.3. *Adapting buildings to the effects of climate change*

Buildings must integrate the climate issues of adaptation to climate change and, through the energy transition, the reduction of greenhouse gas emissions (GHG), involving the need to replace fossil fuels with renewable energy. The building sector is responsible for 44% of national energy consumption and a quarter of greenhouse gas emissions. Heating alone accounts for 70% of the building sector's consumption. In total, the building sector emits more than 120 million tons of carbon dioxide each year, or nearly a quarter of national emissions. Published in 2015, the Energy Transition Law for Green Growth (LTECV, *loi de transition énergétique pour la croissance verte*) sets medium- and long-term quantified objectives to define France's energy and climate evolution (CEREMA 2016). These include reducing greenhouse gas emissions by 40% by 2030 compared to the 1990 baseline, reducing fossil fuel consumption by 30% by 2030 compared to the 2012 baseline, reducing the share of nuclear power to 50% of electricity production by 2025 and increasing the share of renewable energy to 32% of final energy consumption by 2030 and 40% of electricity production, to multiply the quantity of renewable and recovered heat and cold delivered by the heat and cold networks by five by 2030. The law also includes concrete measures in the areas of buildings, transport, the circular economy, renewable energy and nuclear safety. It sets a target of 500,000 home renovations per year, starting in 2017. Its goal is to renovate the entire housing stock to make it "low-energy buildings" or equivalent by 2050, and, in particular, to renovate private residential housing whose primary energy consumption is greater than 330 kWh per m^2 per year, which will have to be renovated by 2025.

The implementation of the law is based on "territorial energy renovation platforms". These platforms are supported by local authorities that are part of a Territorial Climate Energy Plan (PCET, *Plan Climat Energie Territorial*), which has now become a Territorial Climate Air-Energy Plan (PCAET, *Plan Climat Air-Energie Territorial*), and whose goal is to reduce greenhouse gas emissions in order to limit the impact of climate change, in conjunction with their Local Housing Plan (PLH, *Plan local de l'habitat*). The platforms work with individuals, communities and professionals to provide the technical, financial, tax and regulatory information needed to develop renovation projects. These platforms advise consumers, according to their needs, directing them to competent professionals through the PRIS (*Point Rénovation Info Service*, developed under the Housing Energy Renovation Plan or PREH, *Plan de Rénovation Énergétique de l'Habitat*) network. The law introduces new obligations, such as: during a facade renovation or a roof repair, thermal insulation of the treated elements is mandatory; a major renovation must include the installation of equipment allowing the monitoring and control of energy consumption; the development of new habitable rooms in existing residential buildings must also improve their energy performance. Finally, renewable energy supply solutions must be studied. Article 8 of the Energy Transition Act requires new buildings owned by the state, public institutions and local authorities to be exemplary in terms of energy and environmental performance; to achieve, whenever possible, a performance known as "positive energy buildings", producing more energy than they consume, as well as a high environmental performance. By 2050, renovated buildings must not exceed a primary energy consumption of 80 kWh/m^2 per year, a value that can be adjusted according to the geographical context. In 2012, energy-intensive buildings represented nearly one-third of the housing stock.

Other modes of adaptation to climate change are also to be considered, particularly in the case of heat waves, especially in cities where the urban heat island leads to the greening of housing (roofs, façades, etc.). These essential changes are generally part of new developments carried out at the neighborhood level in particular.

7.4. Adapting the neighborhood and the city

The adaptation of housing is often thought within the framework of a district, and more rarely of a city. Eco-neighborhoods serve as a model, integrating housing that limits the use of fossil fuels, green spaces that are useful for biodiversity and effective against the urban heat island, adapted risk and crisis management and so on, but the change of scale from a neighborhood to a city, and, even more so, to the increasingly numerous megacities, remains very difficult. However, new urban planning methods are better adapted to the expectations of a sustainable and resilient city.

7.4.1. *Flood-safe neighborhoods*

The adaptation of building types mentioned above generally falls within the framework of neighborhoods that experience risks, designed to allow living with and when crises occur and to maintain the population in its living space. This is a matter of specific urban planning choices (CEREMA 2015; Rode et al. 2017, 2018).

7.4.1.1. *Amphibious and floating habitat districts*

Since prehistoric times, humans have lived on the water for safety reasons, then later for economic reasons (trade) or tourism. Nowadays, living on the water can be a response to the coastal risks linked to the rise in sea level and the effects of storms and the risks of flooding along rivers. In Amsterdam, for example, as early as 1996, the municipality decided to build the Ijburg district to the east of the city, on the banks of Lake Ijmeer, on artificial islands. Initially consisting of two artificial islands, this new district has been greatly expanded due to high demand. It is an experimental district, based on the construction of floating houses connected to the dike by pontoons. Each house is installed on a submerged concrete tank that is moored to steel poles anchored to the bottom of the water, and these poles allow the houses to follow the water level. In Maasbommel, a village located southeast of Amsterdam on the Meuse River, 46 amphibious houses have been built, adapted to the rising water level. These colored houses, built in wood and aluminum, rest on a hollow concrete raft. They are moored to metal pillars 4 meters high, which contain the flexible tubes of the electrical network, running water and the connection to the sewer. This type of housing entails an additional cost of 15 to 20%, compared to a traditional construction of the same size.

Among the responses to climate change and its effects, particularly the rise in sea level, the example of Rotterdam is outstanding. The city wants to be "climate proof" from 2035, meaning developing floating constructions that can be moved according to need. Thus, the Floating Pavilion, which houses the National Water Center, was designed as a pilot project for floating architecture, the first step towards a floating urbanization, without sliding pillars. The base of this structure of almost 1,000 m^2 is composed of layers of expanded polystyrene (EPS, which contains 98% air), in which layers of concrete have been incorporated to combine buoyancy and stability. Since 2005, the Dutch government has given permission to build "water houses" (floating and amphibious) in 15 areas of the country. However, their implementation raises the question of land, so, in the Netherlands, before becoming the owner of a floating house, it was necessary to be the owner of a plot of water which is constructible; the Dutch government considering the development of floating houses as a necessity has facilitated the purchase of water parcels on which to establish a house. The owners buy the site for a period of 99 years. The costs of such constructions are high, which is an obvious brake to a wide diffusion.

In France, the status of floating homes remains complex; a floating home is considered as an atypical type of craft that does not have its own laws. The establishment of floating homes is impossible on the natural public maritime domain, apart from any activity that falls within the framework of a public service mission. Nevertheless, coastal intercommunities could perhaps implement a new type of procedure allowing the establishment of floating homes on their territory through the Élan law (2018), which creates two new mechanisms: large-scale urban planning operations (GOU, *grandes opérations d'urbanisme*) and development partnership project contracts (PPA) resulting from an agreement with the state. The PPAs, which are contractual commitments, facilitate the conduct and completion of a development operation. The specific tools provided for by the law are the amicable transfer by the state of land in its private domain to a public establishment for intermunicipal cooperation (EPCI); the possible use of the state's public development establishments to carry out studies prior to the preparation of a PPA, even outside their perimeter; and the possibility of defining a perimeter for a large-scale urban planning operation (GOU) to facilitate carrying out development operations. The public river domain can be leased to private individuals or for commercial purposes through temporary, precarious and revocable occupancy agreements issued by public managers (VNF, *Voies navigables de France*, *Ports de Paris*, CNR, *Compagnie nationale du Rhône*, local authorities, particularly regions, etc.) or private concessionaires (private ports, CCI, *Chambre de Commerce et de l'Industrie*). To build a floating home, we need an occupation permit (COT, *Convention d'Occupation Temporaire*), issued by VNF, which allocates a site in the areas previously defined for this purpose. The municipalities are asked to create sites for stationary boats and floating homes. The delimitation of the zone and its specificity (housing, activity, etc.), the works to be carried out and the services to be provided (household waste, water, electricity, etc.) are the responsibility of the waterway manager, after consultation with the municipalities. The owner of the floating home must pay an annual fee. A floating home on private property can be set up in compliance with town planning regulations (Riou 2019). In this case, the owner of the floating home can also be the owner of the land (lakes, ponds). However, in France, these habitats are still quite rare and pose many problems, particularly in terms of urban planning law (Bréard 2018).

7.4.1.2. *Neighborhoods with water transparency*

Breaking with the often-defended idea of moving people away from dangerous areas, a new strategy now envisages the city "with" risk, rather than "against" it. From now on, the link between the city and water is emphasized. "There is hardly a river town that does not have projects, modest or ambitious, to create new urban spaces along the river [...]. The renewed union of cities and their rivers is based on the return in force of a new water culture..." (Allaman 2003). The term resilient urbanism in flood-prone areas is now ubiquitous in urban discourses and strategy documents (Quénault 2014). Resilient urbanism is referred to as "thinking about urban planning so that the area can accommodate water and that buildings and their functioning (services, networks, etc.) are affected as little as possible" (Lesquel 2012). It is now the city that must be adapted to floods; it is no longer the water that must be controlled to allow urban development. Resilient urbanism "seeks to characterize according to which urban, technical, morphological, programmatic and social configurations" it is possible to "design the resilience of a flood-prone urban neighborhood" (Barroca and Serre 2012). This urbanism relies on urban and civil engineering, and especially hydraulic engineering, techniques to predict and characterize flooding on a site-specific basis (Dégardin et al. 1999; Dégardin 2002). In this new approach, blocks or neighborhoods cannot block the water. In the flood-prone Matra district in Romorantin-Lanthenay, a real pilot district in this field (Lacombe et al. 2016), the flood-prone character of the sector was integrated from the design stage of the project in the types of construction and their implantation. Indeed, the space occupied by the former Matra factories located along the river is floodable. In order to transform this industrial heritage into a space for living and leisure, it was necessary to reconcile the weight of history with the constraints of the environment. The overall plan follows the riverbed, and the buildings located near the water adapt to the design of the watercourse. The buildings are designed in order to reduce the impact of flooding. The different buildings are designed to remain accessible despite the rise in water levels, the housing levels are located above the PHEC and the parking lots are located below. Explaining and preparing the population for floods was also necessary. The neighborhood was flooded during the Sauldre River flood of June 2016, which tested the development choices made. Vehicles had been preventively evacuated, and the population was able to stay in their homes and continue living in the flooded neighborhood. The costs of a resilient development operation are diverse: financial, political and so on, and we sometimes speak of the additional costs associated with resilience. Other aspects concern the sale of such properties, will their specificity be a selling point in the future? The political costs are not insignificant; opposition to such constructions should not be overlooked and may contribute to slowing down this type of practice (Lacombe et al. 2016).

These hydraulic transparencies raise the question of the quality of public spaces in the heart of new districts. Their contribution to the urban character of the area, which is a major issue for the inhabitants, is not necessarily self-evident. If these spaces are considered only in terms of their hydraulic function, and not also in terms of their daily function as public spaces, it will be difficult for the inhabitants to appropriate these places, which will certainly be adapted to flooding but less to the creation of social links. A resilient urbanity requires a close association of these different aspects, which is far from always being the case.

7.4.1.3. *Slab neighborhoods and their limits*

Elevation allows building in flood-prone areas (CEPRI 2015; Rode and Gralepois 2017), to put housing, critical infrastructure and traffic axes out of water. Construction on a slab was widely used on the scale of more or less vast districts in the second half of the 20th century, as evidenced by the development of the flood-prone district of Angers, the Saint Serge district. This area, a wetland filled in the 19th century to set up a goods station, industrial and commercial activities (national interest market in 1963) became a Urban Development Zone (ZAC, *Zone d'Aménagement Concerté*) in 1991. The ZAC includes a business center, a university center, leisure facilities and housing. Completed in 2000, there is also a park in the heart of the ZAC, built on a slab above the public parking lot. This park also serves as a safety platform in the case of flooding of the Maine River. The first level of the houses is located 50 cm above the level of the PHEC, corresponding to the 1995 flood. The neighborhood thus created required numerous stairs and ramps, while the first floors are devoted to parking lots which water cannot, in principle, penetrate because they are elevated, but in the event of these parking lots flooding, the water stagnates there for a long time. This area is not conducive to the establishment of businesses and the traffic involving stairs and ramps is a hindrance to an active urban life. The results of the operation are therefore questionable, which explains the more recent developments in the flood-prone areas of Angers.

Despite the disadvantages mentioned, other examples of construction on slabs have been carried out recently, such as the Eure-Saint Nicolas district in Le Havre, an interface space between the port and the city, a witness to the economic, maritime and port history of Le Havre, the industrial and port wastelands around the basins and the former docks have been redeveloped. Nearly 2,000 housing units were thus established between 2008 and 2015, more than 80% in the Eure district alone and 11% in the Arcole/Brindeau district (Orillard et al. 2018). This district, like a significant part of the city of Le Havre, is affected by the risk of flooding by marine submersion, runoff and the rising water table. A specification taking risk into account compensated for the absence of a PPR when launching the new development. The design and construction of the neighborhood is based on the elevation of habitable floors and non-habitable basements.

7.4.1.4. *Urban planning in flood-prone areas: embankments and excavations*

In Angers, the Berges de Maine mission was created in 2008, with the aim of integrating the city and its river. The developments along the Maine River are evolving from slab constructions, whose limits have been underlined, to another form of development compatible with the floodable nature of the space. The SCOT Pays-Loire Angers, approved in 2011, takes into account the Berges de Maine project, and the Saint-Serge sector is designated in the General Guidelines Document (DOG, *Document d'Orientations Générales*) as a major strategic site intended to increase the attractiveness of downtown Angers (housing, facilities, services, etc.). The city wished to abandon slab-based urban planning and to protect part of the developable area from water by means of an embankment/excavation system. Such a project is the result of the water law which "to preserve the free flow of water and the expansion field of floods, can impose to the developments in major bed to compensate their right of way in volume, in surface and in altimetric dimension of functioning"; the implantation of embankments implying a compensation "in volume, dimension for dimension". In flood-prone areas, embankment can only be authorized if its hydraulic impact is compensated by new flood expansion capacities. In order to do this, a largely submersible park of 20 ha in the Saint-Serge district, the Parc des Confluences, would have to be created, implying an important movement of excavation/re-embankment, with 100,000 m^3 of earth to be excavated in the park to allow for the development of Saint-Serge. A harbor and two islands (one wild and one for leisure) that are submersible in the case of a flood should be established. The excavated material recovered by the creation of the park would therefore allow the level of the urban plateau to be raised 50 cm above the PHEC. The movements of excavations and embankments should allow the construction of new houses and activities on the filled areas, in order to reduce vulnerability. The objective is to secure the neighborhood by facilitating the flow of water, an urban space free of water, would allow the implementation of constructions whose first floor would be directly accessible from the public space. This project, which adds to the control of risk, the preservation of the environment (green and blue network) and soft displacements, can be considered as resilient.

7.4.2. *Urban technical networks and resilient critical infrastructures*

Making infrastructures robust to hazards for increased urban resilience must be a major concern of risk and urban sustainability stakeholders (Lhomme et al. 2010, 2012; Toubin et al. 2012). When subject to crisis situations, the vital functions provided by critical infrastructures (IC, *Infrastructures Critiques*) must be restored or adapted as quickly as possible to return to a minimum required level of functioning. This capacity to maintain or re-establish a minimum level of functioning despite disruptions refers to the concept of resilience, not only of the critical infrastructures themselves but, more broadly, of the urban territory concerned. Knowledge of the resilience capacity of critical infrastructures therefore requires an interest in both the safety of users and residents, and the societal and economic consequences that a lasting interruption of services may cause. Critical infrastructures defined as "activities of vital importance (AIV, *Activités d'importance vitales*) (Art. L. 1332-2 of the Defense Code) cover 12 sectors grouped into four dominant sectors by the General Secretariat for Defense and National Security (SGDSN): human (health, water management and food activities); regal (State military, civil and judicial activities); economic (transportation, energy and financial activities) and technological (communication, space, research and industry activities)". The interweaving of these different domains explains the difficulty of considering various damage scenarios and their consequences. In addition, the vulnerability of critical infrastructures and of the territory concerned can result from a domino effect, thus following an earthquake, a flood or any other event (exogenous or endogenous), several critical infrastructures can be affected. The treatment of these malfunctions implies both reducing the material damage, to allow a functioning in a degraded mode and a fast resumption of service. To achieve this, it is necessary to act in fields other than the technical one and to involve administrations, economic and political actors and territories of various sizes. The resulting hypercomplexity is responsible for uncertainty, which hinders anticipation, limits planning and complicates conduct in the event of a crisis. Thus, as Emmanuelle Moatty and Magali Reghezza-Zitt show, in the case of "sanitation critical infrastructures, the restoration of equipment and infrastructure for the collection, transport and treatment of effluents depends in particular on the restoration of roads, as well as of electricity and telecommunications" (Moatty and Reghezza-Zitt 2018). Technical resilience is directly linked to, and even directly dependent on, organizational resilience (the ability to manage facilities, maintain key functions and make decisions to maintain/improve the situation during the event), the behavioral aspects of the populations subjected to the crisis, and economic data. Thus, the action of health services during the coronavirus pandemic in 2020–2021 (organizational aspect) can be jeopardized by the displacement of part of the population during vacation periods (human aspect), which contributes to a faster circulation of the virus. On the other hand, lockdown has led to great economic difficulties

7.4.3. *Responses to climate risks – the urban heat island, "green" neighborhoods*

Flooding can be reduced by developing or redeveloping densely built-up areas in the form of neighborhoods or "sponge cities". The vegetation of concrete spaces allows the absorption and slowing down of water. In Rotterdam, for example, in addition to the development of abandoned port areas, the management of water in the city to cope with the risk of flooding explains the transformation of several public squares into "water squares", that is, squares that are temporarily floodable. Thus, at least part of the water is retained during heavy rainfall. Retention basins collect the water from the roofs of buildings and from the street. The water remains in the basins for a maximum of 48 hours before being released to the sewer system. Some of it can also be directed to green spaces. Rotterdam built the first floodable multifunctional square, the Benthemplein, in 2013. The three basins that make up the Benthemplein are used for sports and leisure activities outside the flood period (basketball court, amphitheater and skatepark). The green solutions concerning the habitat as well as the entire urban space are also useful for reducing summer heat, to limit the extent of the urban heat island (UHI), aggravated by the nature of the materials constituting the buildings, the traffic lanes, the essentially inorganic use of the land, the specificities of the urban morphology and human activities. The UHI has various impacts on health and the well-being of the inhabitants; it also has consequences on energy consumption with the strong use of air conditioners, not to mention the consequences on biodiversity. Communities use various solutions to reduce the UHI: "green" solutions, linked to nature and plants, "blue" solutions linked to the presence of water and "gray" solutions linked to urban materials. The creation of gardens and/or parks, the densification of vegetation in squares, along roads and even on the building itself is a first response. The presence of water, fountains, water jets, watering of pavements, installation of water retention paving stones, existence of ditches to collect runoff water, underground storage basins of rainwater for surface watering and so on can also be associated with light-colored pavements and/or buildings. To this should be added the specificities of urbanization, urban forms, which influence the urban heat island effect in relation to the multiplication of reflection surfaces and the disruption of air flows.

7.4.4. *Eco-neighborhoods, resilient neighborhoods?*

Eco-neighborhoods, or *écoquartiers* in France, the emergence of a global approach to sustainable development in urban planning, integrate responses to risks and crises, an ex ante resilience. Eco-neighborhoods should, through the urban planning choices they underpin, reduce the scale of future crises by drawing on feedback from past crises; they should therefore provide examples of resilience. One of the objectives of eco-neighborhoods is to implement a project approach that responds to the global challenges of the planet: climate change, degradation of biodiversity, careless use of resources, as well as local issues (existence of natural and technological risks, etc.). The urban planning of eco-neighborhoods must contribute to improving the quality of life of the inhabitants and users, as well as to the sustainability of the city to which the eco-neighborhood belongs. Thanks to a strong political will and a strong involvement of the inhabitants, eco-neighborhoods are based on a new way of thinking and acting, established on "systemic" approaches (water cycle, energy, circular economy, etc.), breaking with strictly sectorial planning practices. Eco-neighborhoods are directly involved in the fight against climate change; they are also involved in water saving, waste reduction, biodiversity protection and the reduction of urbanized areas. The specificities of the urban spaces of eco-neighborhoods carry values (for future generations in particular) and beliefs in the capacity of urban forms to shape societies. However, past examples, large housing estates and new towns that were also supposed to be "the laboratory in which the social structures of tomorrow are developed" have not given the expected results.

In France, the *ÉcoQuartier* label, carried since 2012 by the Ministry of Housing, was renewed in 2016, with the aim of moving from the *ÉcoQuartier* "laboratory of experiments" to the *ÉcoQuartier* model of sustainable development of tomorrow and established on 20 commitments, divided into four areas: approach and process, living environment and uses, territorial development, environment and climate. Among these 20 commitments, we will highlight the three that directly take crises into account, commitment 8: to ensure a safe living environment that integrates the major health issues, particularly air quality; commitment 16: to produce urban planning that anticipates and adapts to risks and climate change; and commitment 17, to aim for energy conservation and diversification of resources to the benefit of renewable and recovered energy.

Eco-neighborhoods are largely part of recent energy policies that aim to reduce greenhouse gas emissions and promote green energies (biogas, etc.), mainly through a low-consumption housing policy, of which the Bonne eco-neighborhood in Grenoble was one of the pioneers. Today, the even more ambitious objectives of RT 2020 call for the construction of positive energy housing or buildings that produce more energy than they consume, and passive houses, that use very little energy and recycle the energy they produce (photovoltaic panels, etc.). In addition to bringing housing up to standard, an effective energy policy requires considering new forms of mobility, so-called "soft" and "short-distance" transport (walking, cycling, public transport), which apply within the eco-neighborhood and to its connection with the city in which it is located. Similarly, waste collection must be carried out from voluntary drop-off points to limit the need for dumpsters. These changes in practices also have the advantage of reducing local air pollution and noise. In order to be as effective as possible, these energy innovations, whether technical or the result of the diversification of energies, must be part of organizational innovations.

Some eco-neighborhoods are trying to adapt to risks, such as flooding. This is the case of the "Rives du Bohrie", on the outskirts of Strasbourg, on the left bank of the Ill, installed in the flood plain. This eco-neighborhood is a model of a resilient district, which by anticipating, from its design stage, that it must reduce the consequences of flooding, but its practices are safe from criticism, since the buildings are installed on platforms connected by pontoons.

The urban planning of eco-neighborhoods, which must adapt to the effects of climate change and risks, must also contribute to reducing energy and resource consumption and to protecting biodiversity. Are eco-neighborhoods always as "virtuous" as their objectives dictate? Many of them are attractive spaces for the wealthy middle classes, to the detriment of the poorest, because of the cost of the buildings, which is sometimes higher than that of nearby neighborhoods, and often also because of an ideological choice. Many of their inhabitants are well informed about ecological issues and the effects of climate change, but this information does not always justify adapted practices, particularly in terms of daily transportation. “There is therefore a tension between awareness and action” (La Branche 2012) that limits the purpose of these neighborhoods in terms of risk and crisis reduction.

7.4.5. *From the eco-neighborhood to the “smart city”*

The transition from a sustainable neighborhood to a sustainable city is not easy to implement. Many cities are trying to deal with natural and/or technological risks, but in many cases, this policy only concerns one district of the agglomeration, as shown by the management of East London. With the TE2100 (Thames Estuary 2100) project, “the British Environment Agency envisages a risk territory that draws its identity from its resilience, that is, its ability to live with the river”. The TE2100 project is a long-term risk adaptation plan rather than a risk reduction plan. We can therefore consider that the East London district is characterized by an intrinsic state of resilience foreseen at the time of its realization and maintained, in particular, by the culture of risk (Beucher 2008, 2012), but is what is true for a district true for the entire city? Can we say that the London metropolitan area is resilient in its entirety? In other cases, the emphasis is placed on greening the city and developing soft traffic (bicycles, etc.), but the sustainable city is still largely utopian. Does the “smart city” make it possible to envisage a sustainable and resilient city? Born in the United States, the expressions “smart city”, “digital city”, “connected city” and “green city”, or sometimes “intelligent city”, cover urban innovation strategies based on information and communication technologies. These technologies are considered to respond to problems that are related to urbanization, resource management and the effects of climate change. In other words, “smart cities” are intended to be resilient insofar as their objectives are the reduction of risks at the source and the best way to manage crises. The European Parliament defines a “smart city” as “a city that seeks to address issues of public interest through digital solutions, enabled by a partnership with multiple stakeholders”. It is a new urban model that is designed to improve the living conditions of citizens. The smart city must be collaborative. The various services offered by the city need to be connected to each other. This, from the centralization of information, allows a simplified and automated management of the city. However, a smart city is still more of a process than a goal. Most of the policies that attempt to implement a smart city insist on the importance and the place given to the citizens in this context, the users-citizens, who must be placed at the heart of the process. Thus, Copenhagen smart city, green and intelligent, should be the first carbon neutral city by 2025. It has more than 400 kilometers of bicycle paths used every day by more than 35% of its population.

The city is therefore involved in the fight against the effects of climate change through (soft) transport modes, the use of renewable energies, connected street furniture and the use of open data (digital data of public or private origin that is free for users to access and use). Connecting street furniture can help reduce a city's environmental footprint (energy savings, intelligent lighting, reduction of greenhouse gases, connected garbage cans with reduced collection times). By monitoring air quality live in the city, it is possible to divert traffic from certain roads if pollution is high and offer parking spaces. These different data need to be widely shared by the population and by companies, which must also participate in their collection in order to improve the quality of life by reducing the risks for the whole population. Many examples of "smart" practices in cities exist in France, and many local authorities are implementing smart energy networks. For example, the city of Paris is planning on transforming schoolyards into cooling islands and to map its basements (3D mapping) in order to recover heat from the sewers in winter. The effort is global. Along with Quito and Rotterdam, Paris is part of a network of a hundred resilient cities around the world.

In 2018, more than 74 digital citizen participation tools for territories were identified. Yet part of the population is suspicious of the "technological" city. To try to reassure these actors, the implementation of a regulation for data protection is essential, and it recommends choosing digital platforms that allow the protection of personal data. Platforms must comply with the obligations of the French National Commission for Information Technology and Liberties (CNIL, *Commission nationale de l'informatique et des libertés*). They must comply with the General Data Protection Regulation (data anonymization, etc.). The local authority must be able to know and monitor the use of personal data managed by the platform. Can the smart city increase urban resilience, especially in the face of various types of crises? It should probably contribute to this, in the more or less long term, in terms of energy, resources (water), biodiversity and the health of populations (reduction of pollution).

In many cases, the desire to transform the urban fabric is based on predominantly ecological arguments. The city is thus treated as an ecosystem whose metabolism must be modified by considering the inputs and outputs of materials and energy. The risk is therefore reducing the sustainable city to purely ecological aspects. This urban utopia refers to the belief, often denied in the past, of the transformation of society through new urban specificities. The social dimension and inequalities in particular are largely forgotten, which reduces the interest of the sustainable city and its possible resilience (Levy 2009).

7.5. Strategic retreat

"Strategic retreat" is the extreme solution, which amounts to displacing populations that are located in particularly dangerous sectors.

After a disaster (flood, earthquake, etc.), the question may arise whether the affected area or even the city should be rebuilt. During the crisis, many people sometimes leave the disaster area, where they lost their homes, as was the case in New Orleans. These populations are sometimes called "environmental refugees", defined as people who have had to leave their traditional habitat, temporarily or permanently, following an environmental disaster that has endangered their existence or changed their living conditions. In some cases, however, the departure of populations is the result of political choices that are made necessary by the dangerousness of certain areas. In France, this is what is known, in the narrow sense, as "strategic retreat", which consists of moving homes and infrastructures installed on a coastline, along a river, or even in other dangerous places (subject to technological hazards) in order to protect vulnerable populations and facilities. This displacement is necessary when the life of the populations is at stake and when the cost of protection works and their maintenance is higher than the value of the endangered goods. Retreat may be the consequence of the crisis or disaster, as was the case on the Atlantic coast after the passage of Xynthia in 2010, sometimes this retreat precedes the disaster because of the knowledge of significant hazards (Verspieren 2018). Indeed, the knowledge of the coastal dynamics and the history of the latter allow us to recognize the places affected by the recession of cliffs and dune strips. Retreat is then considered within the framework of the SOLTER project (*solidarités territoriales et stratégies pour la résilience du littoral à la submersion marine*, territorial solidarities and strategies for coastal resilience to marine submersion), which aims to study solutions for adaptation to rising water levels; following a call for projects launched by the Ministry of the Environment in 2012, to consider anticipated relocation in seven municipalities (Ault, Hyères, Lacanau, La Teste de Bush, Labenne, Petit Bourg (Guadeloupe) and Vias). Current French law provides few solutions for considering the "deconstruction" of existing buildings. During the Xynthia storm, the urgent solutions taken (the demolition of several hundred houses) cannot be transposed to all French coastlines threatened. By extending the benefit of the "Barnier" fund to damage due to marine submersions, it was possible to expropriate the occupants (of the "black zones"), with compensation allowing "the replacement of expropriated property", not taking into account the devaluation of the property due to its exposure to risk. Such a generous system, which is very favorable to owners who would probably not have found a way to sell their property given the dangerous nature of the area, cannot be widely used, particularly because of its cost. The legal tools that impose the abandonment of threatened buildings by their owners can be summarized as amicable purchase (L651-3 Environment Code) or expropriation (L561-1 Environment Code) with compensation. If we "let the sea rise", which leads to taking note of the submersion of a property by the highest waters, this property will automatically be incorporated into the public maritime domain, without any compensation (Article L2111-5 general code of the property of public persons). Thus, new legal approaches envisage staggering the abandonment procedures for built-up property over time and modulating compensation. The abandonment of coastal property over long periods (one or two generations) should reduce the shock of the sudden loss of property, as the owner accepts the gradual abandonment of their property, on which they would no longer invest, either financially or emotionally, knowing that it is doomed. Staggering also allows the community to spread out the eventual compensation. Abandonment could begin over an initial period of 30 years, with the owner first losing the *abusus* of their property, which prevents them from reselling it or passing it on to their heirs; the property may be occupied by its owners, their successors or their tenants as long as this occupation does not endanger the lives of the occupants. However, the property would be inalienable. Over a second 30-year period, the owner would lose the *fructus*, it would no longer be possible for them to rent the property, but it would still be possible for them to inhabit it as an owner-occupant; over the final 30-year period, the owner would finally lose the use (*usus*) *of* their property (Lambert 2015). The compensation itself in this proposal requires reflection. The MAReL method proposes criteria of social equity involving the qualification of permanent or secondary residence and the occupancy rate of the building, the tax capacity of the owner, the use of the building that is essential to the profession (residence of fishermen or oyster farmers); "safeguarding" rules should allow us to verify the date of acquisition or duration of occupancy of the compensated property, to avoid speculative acquisitions. Thus, this version of solidarity could be proposed that is, in a way, differentiated according to the economic vulnerability of the owner-occupier (do they have other properties? can they easily reinvest in another dwelling?) (Lambert 2015).

On the Atlantic coast in Lacanau, coastal erosion is threatening 1,200 homes and a hundred businesses. In the long term, a "strategic withdrawal" is envisaged for these properties, that is, 20% of the secondary residences and most of the commercial activity. Criel-sur-Mer (Seine-Maritime, Normandy) offers another emblematic case, as part of the commune is located on high chalky cliffs that are rapidly eroding. Thus, between 1995 and 1998, the cliff retreated by 1.3 to 3.3 meters/year in some places and several houses were in danger near the top of the cliff. As early as 1996, the mayor of Criel led negotiations with the state to proceed with the repurchase of 10 houses located at the edge of the cliff. To justify this measure, the cost of protection was compared to the cost of compensation in case of expropriation and the cost of destroying the houses. In the end, these houses were not expropriated, but bought out of court using the Barnier fund. Once the buildings were destroyed, the land was restored and the space was given to the *Conservatoire du Littoral*. In Wimereux and Equihen-plage, about 15 properties that were exposed to the recession of the cliffs were also bought and then destroyed using the Barnier fund (Mineo-Kleiner and Meur-Ferec 2016). In Hyères, in the Provence-Alpes-Côte d'Azur region, the municipality has a relocation project. Its coastline is suffering from the retreat of the sandy coast, marine submersion and river flooding. The municipality is therefore planning to move a road that runs along the seafront, and on its site, a soft traffic lane will be built. The departmental road to be relocated serves a mixed civilian and military airport, housing estates, tourist facilities (campsites, hotels, etc.) and farms located on the alluvial plain of the Ceinturon. In Ault, in Picardy, a village is located on chalky cliffs that have been retreating for a long time. Considerable work to stabilize the cliffs has heavily indebted the municipality, with no effective result. It is therefore necessary to relocate houses, hotels, restaurants and bars and the community hall. The relocation file is carried by the municipality of Ault and the mixed syndicate in charge of the management of the coastline. The initial project is to gradually relocate the assets. The enhancement of the cliff edge by the development of a cornice with mobile urban equipment could help revitalize the city (Mineo-Kleiner et al. 2016).

With regard to technological risks, expropriation and relinquishment concern the most dangerous zones defined in the PPRT, with expropriation corresponding to areas where the urgency of the situation requires a response from public authorities. In the case of relinquishment, the community is obliged to buy back the property relinquished by its owner in dangerous areas. This is a right that is open to local authorities – not an obligation like expropriation – allowing the acquisition of a property that is put up for sale. Nevertheless, expropriation and relinquishment give rise to a financing agreement between the state, the local authorities and the industrialists. In contrast, preemption is left to the exclusive financing of local authorities in charge of urban planning (Gralepois 2012). The right of preemption, which is exercised over the entire perimeter of the PPRT, is a possibility offered to local authorities to intervene on the land market to acquire, at their own expense, any real estate within the perimeter of the PPRT.

The possibility for local authorities to acquire land classified within the perimeter of the PPRTs facilitates the formulation of development scenarios for areas that are currently declared unbuildable, but which may later be the target of development projects. This possibility of acting on the urban area is analyzed as a "total reversal of perspective" of the logic of intervention by public authorities, which installs "artificial spaces" or "emptiness" between the factories and the city (Bonnaud et al. 2005). Thus, certain dangerous and unbuildable territories can also represent land reserves or spaces that allow for the expansion of factories. If the PPRTs lead to a certain control of the hazardous area defined by state services, the implementation of urban planning measures is the responsibility of local authorities (the commune or the intermunicipality).

In many cases, the assets to be displaced are considered as needing to be relocated, but the choice of the relocation territory is generally not unanimous. Often, for municipal elected officials, the relocation must take place within their municipality, while, for other actors, the relocation could be done on an intermunicipal scale, or even on the scale of a territorial coherence scheme (SCoT), the contours of which seem to be the most relevant for developing a relocation project.

7.6. Conclusion

Rebuilding after a crisis or renewing urban space by integrating the potential dangers and vulnerabilities of the territorial system is a major challenge. The issue is all the more important as cities, which have been expanding rapidly over the last few decades, are faced with crises, not only localized but also global crises (impacts of climate change, epidemic crises, etc.), the solution to which is partly beyond the control of local territories and actors, and which therefore require modes of regulation on an international scale.

However, good governance, whether local, national or international, to meet the challenges of future crises is not enough. Social trust also counts. The economic component must also be considered. Thus, it is the entire territorial system that is being questioned.

The challenge is all the more important because it is a matter of introducing stability into open and dynamic territories, which are intrinsically unstable by nature. Limiting the scale of crises and preventing the most harmful consequences will therefore require offensive and intelligent measures and the search for a balance between system dynamism and security. This balance will require even greater investment in prevention, preparation and adaptation, the cost of which will always be negligible compared to the economic and human losses that crises can cause.

References

Ahuirv, G. (2016). Gestion de crise : l'appropriation territoriale à l'épreuve des faits. Geography Thesis, Université Panthéon-Sorbonne, Paris.

Aldrich, D.P. (2012). *Building Resilience: Social Capital in Post-Disaster Recovery*. University of Chicago Press, Chicago.

Alexander, D.E. (1997). The study of natural disasters some reflections on a changing field of knowledge. *Disasters*, 21(4).

Alexander, D.E. (2010). The L'Aquila earthquake of 6 April 2009 and Italian government policy on disaster response. *Journal of Natural Resources Policy Research*, 2(4).

Allaman, M. (2003). L'odyssée des villes et de leur fleuve. *Diagonal*, 163.

Allinne, C. (2007). Les villes romaines face aux inondations. La place des données archéologiques dans l'étude des risques fluviaux. *Geoarchéologie, Revue Geomorphologie*, 13(1).

Altez, R. and Revet, S. (2005). *Contar los muertos para contar la muerte : discusión en torno al número de fallecidos en la tragedia de 1999 en el estado Vargas – Venezuela*. Revista Geográfica Venezolana, Numero especial.

Aluson, G. and Zelikow, P. (1999). *Essence of Decision*, 2nd edition. Longman, New York.

Anderson, M. (1989). *Rising from the Ashes*. Routledge, London.

Angeli-Aguiton, S., Cabane, L., Cornilleau, L. (2019). Politiques de la "mise en crise". *Critique internationale*, 4(5).

Arboit, G. (2007). *De l'utilisation médiatique des catastrophes. L'exemple du tsunami de la Saint-Etienne 2004*. Annuaire Français de Relations Internationales, VII, Bruylant.

Aublet, B. (2014). L'action en situation d'urgence : facteurs d'efficacité dans la gestion du réseau routier en cas de crues rapides : etude appliquée au département du Gard. Geography PhD Thesis, Université de Grenoble.

Augé, M. (1994). *Pour une anthropologie des mondes contemporains*. Aubier, Paris.

Territorial Crisis Management,
coordinated by Richard LAGANIER and Yyvette VEYRET.

Avila, L.A. and Rappaport, E.N. (1996). Atlantic hurricane season of 1994. *Monthly Weather Review*, 124(7).

Baird, A., Keefe, P.O., Wesgate, K., Wisner, B. (1975). *Toward an Explanation and Reduction of Disaster Proneness*. Bradford University, Distaser Research Unit, Bradford.

Ballais, J.-L. and Ségura, P. (1995). La catastrophe de Vaison- la-Romaine (22 septembre 1992) : une inondation dans un géosystème anthropisé. *120ème congrès national des sociétés historiques et scientifiques*, Aix-en-Provence.

Bankoff, G., Frerks, G., Hilhorst, D. (2004). *Mapping Vulnerability: Disasters, Development and People*. Earthscan, London.

Barclay, J., Wilkinson, E., White, C.S., Shelton, C., Forster J., Few R., Lorenzoni, I., Woolhouse, G., Jowitt, C., Stone, H. et al. (2019). Historical trajectories of disaster risk in Dominica. *International Journal of Disaster Risk Science*, 10.

Baron-Yellès, N., Goeldener Gianella, L., Velut, S. (2002). *Le littoral. Regards, pratiques et savoirs*. Éditions Ens, Rue d'Ulm, Paris.

Barraqué, B. and Moatty, A. (2020) The French Cat' Nat' system: Post-flood recovery and resilience issues. *Environmental Hazards*, 19(3).

Barroca, S., Serre, D., Youssef, D. (2012). Le concept de résilience à l'épreuve du génie urbain. *Vertigo*, 12(2).

Barus-Michel, J. (2009). Crise(s). *Les Cahiers de Psychologie Politique*, 14.

Barus-Michel, J., Giust-Desprairies, F., Ridel, L. (1996). *Crises : approche psychosociale clinique*. Desclée de Brouwer, Paris.

Basu, M., Ghosh, S., Jana, A., Bandyopadhyay, S., Singh, R. (2017). Resource mapping during a natural disaster: A case study on the 2015 Nepal earthquake. *International Journal of Disaster Risk Reduction*, 24.

Baudry, H. (2009). Médecine et médecins face au tremblement de terre de Lisbonne en 1755. *Dix-huitième siècle*, 41.

Bazin, P. and Olivry, D. (2007). Le conservatoire du littoral : un acteur face au changement climatique. *Revue Science Eaux & Territoires*, Restauration et réhabilitation des zones humides : enjeux, contextes et évaluation, 24.

Beck, U. (2001). *La société du risque. Sur la voie d'une autre modernité société*. Aubier, Paris.

Benitez, F. (2018). Faire face ou vivre avec les catastrophes ? Capacités d'adaptation et capabilités dans les trajectoires de résilience individuelles et territoriales au sein de l'espace Caraïbe. PhD Thesis, Université Paul Valéry, Montpellier.

Benitez, F. (2019). La catastrophe de la Montagne Pelée le 8 mai 1902 en Martinique : Saint-Pierre, une ville résiliente ou un exemple archétypal de bifurcation. *Physio-Géo*, 14(1).

Beraud, H. (2013). Initier la résilience du service de gestion des déchets aux catastrophes naturelles : le cas des territoires urbains de l'inondation. Thesis, Université Paris-Est, Marne-la-Vallée.

Bergeron, H., Borraz, O., Castel, P., Dedieu, F. (2020). *COVID-19 : une crise organisationnelle*. Presses de Sciences Po, Paris.

Berke, P.R. (1995). Natural-hazard reduction and sustainable development: A global assessment. *Journal of Planning Literature*, 9(4).

Berke, P.R., Kartez, J., Wenger, D. (1993). Recovery after disaster: Achieving sustainable development, mitigation and equity. *Disasters*, 17(2).

Berque, A. (2012). Milieu, co-suscitation, désastres naturels et humains. *Ebisu*, 47.

Berque, A. (2018). Pour une architecture paysagère. Conférence 19 juin 2018 [Online]. Available at: http://ecoumene.blogspot.com/2018/06/pour-une-architecture-paysagere.html.

Bertolini, G. (1990). *Le marché des ordures : economie et gestion des déchets ménagers*. L'Harmattan, Paris.

Bethemont, J. (1972). Le thème de l'eau dans la vallée du Rhône. PhD Thesis. Edition Le Feuillet Blanc, Lyon.

Beucher, S. (2008). La gestion des inondations en Angleterre : la mise en place d'un système efficace de gouvernance du risque ? *L'Information géographique*, 12(4).

Beucher, S. (2012). *Renouvellement urbain et développement durable de l'Est londonien : vers la mise en place de territoires résilients au risque d'inondation*? ENS Ulm [Online]. Available at: http://www.geographie.ens.fr/IMG/file/resilience/resilienceulm.pdf.

Binot, A. and Karsenty, A. (2007). La question foncière, les ressources naturelles et l'environnement. *VertigO - la revue électronique en sciences de l'environnement*, Horssérie 4.

Birkland, T.A. (1998). Focusing events, mobilization, and agenda setting. *Journal of Public Policy*, 18(1).

Blaikie, P., Cannon, T., Davis, I., Wisner, B. (1994). *At Risk: Natural Hazards, People's Vulnerability and Disasters*. Routledge, London.

Blin, A. (2005). *Le terrorisme*. Le Cavalier Bleu, Paris.

Bode, B. (1989). *No Bells to Toll. Destruction and Creation in the Andes*. Charles Scribner's Sons, New York.

Boen, T. (2006). Observed reconstruction of houses in Aceh seven months after the Great Sumatra earthquake and Indian Ocean tsunami of December 2004. *Earthquake Spectra*, 22(3_suppl). doi: 10.1193/1.2211369.

Boin, A. and Hart, P. (2006). The crisis approach. In *Handbook of Disaster Research*, Rodriguez, H., Quarantelli, E.L., Dynes, R.R. (eds). Springer, New York.

Boin, A. and Hart, P. (2010). Organizing for effective emergency management: Lessons from research. *Australian Journal of Public Administration*, 69(4).

Boniface, P. (2020). *Géopolitique du COVID-19*. Eyrolles, Paris.

Bonnaud, L. and Martinais, E. (2005). Des usines à la campagne aux villes industrielles. *Développement durable et territoires*, Dossier 4.

Bonnet, J. (ed.) (2012). *Aménagement et développement territorial*. Ellipses, Paris.

Bouchon, S. (2009). L'analyse de la vulnérabilité pour identifier les infrastructures critiques européennes : exemple du système gazier en Europe. In *Risques et environnement : recherches interdisciplinaires sur la vulnérabilité des sociétés*, Becerra, S. and Peltier, A., (eds). L'Harmattan, Paris.

Bouchon, S. (2011). L'identification des infrastructures critiques : Réflexion à partir de l'exemple européen. PhD Thesis, Université Paris Ouest Nanterre La Défense, Nanterre.

Boudières, V. (2007). De la gestion des risques, aux risques de la gestion. *Géocarrefour*, 82(1–2).

Boudjella l, L., Vial, T., Vautier-Larrey, A., Mériaux, P. (2013). Mise en place de la réglementation issue du décret du 11 décembre 2007 pour les digues maritimes des départements côtiers de la Manche et de Bretagne : état d'avancement, exemples marquants et perspectives. *Colloque national digues maritimes et fluviales de protection contre les submersions*, Aix-en Provence.

Bourdieu, P. (2006). Le capital social. Notes provisoires. In *Le Capital Social, Performance, équité et réciprocité*, Antoine B. and Michel L. (eds). La Découverte, Paris.

Bourrelier, P.H., Deneufbourg, G., de Vanssay, B. (2000). *Les catastrophes naturelles, le grand cafouillage*. Editions Osman Eyrolles Multimédia, Paris.

Bravard, J.P. and Clemens, A. (2008). *Le Rhône en 100 questions*. Graie. ZABR (Zone Atelier Bassin du Rhône), Lyon.

Bréard, L. (2018). L'habitat flottant en France : cadre juridique et perspectives. Master's Thesis, CNAM-ESGT, Paris.

Brussee, A.R., Bricker, J.D., De Bruijn, K.M., Verhoeven, G.F. Winsemius, H.C., Jonkman, S.N. (2021). Impact of hydraulic model resolution and loss of life model modification on flood fatality risk estimation: Case study of the Bommelerwaard. *Journal of Flood Risk Management*, 14(3).

Brusset, E., Collin, C., Fusillie, R.C. Spaak, M., Willot, P. (2011). *Etude des interventions post-catastrophe*. AFD, Paris.

Buteau, P., Saint-Eloi, R., Trouillot, L., Acacia, M., Auguste, L. (2010). *Refonder Haïti ?* Mémoires d'encrier, E-book.

Cabane, L. (2015). Les catastrophes : un horizon commun de la globalisation environnementale ? *Natures Sciences Sociétés*, 23(3).

Cabane, L. and Revêt, S. (2015). La cause des catastrophes, concurrences scientifiques et actions politiques dans un monde transnational. *Politix*, 3(111).

Campanella, T.J. (2006). Urban resilience and the recovery of New Orleans. *Journal of the American Planning Association*, 72(2).

Carbiener, R. (1983). Le grand Ried central d'Alsace : écologie et évolution d'une zone humide d'origine fluviale rhénane. *GéoProdig, portail d'information géographique* [Online]. Available at: http://geoprodig.cnrs.fr/items/show/169032.

Cartier, S. and Valette, C.L. (2016). Parasismique, oser éprouver l'acceptabilité sociale. *Vertigo*, 16(1).

CCE (2005). *Livre vert sur un programme européen de protection des infrastructures critiques*. Commission des Communautés Européennes, Brussels.

CEPRI (2010). *Le bâtiment face à l'inondation : diagnostiquer et réduire sa vulnérabilité. Guide méthodologique*. CEPRI édition, Orléans.

CEPRI (2015). *Comment saisir les opérations de renouvellement urbain pour réduire la vulnérabilité des territoires inondables face au risque d'inondation ? – Principes techniques d'aménagement*. CEPRI edition, Orléans.

CEREMA (2015). Villes et territoires résilients. In "*Études et documents*" *de la Délégation au développement durable* (DDD). Commissariat Général au Développement Durable (CGDD n°123 2015), Paris.

CEREMA (2016). *Décryptage de la loi de transition énergétique Fiche n° 03. Bâtiments : que dit la loi de transition énergétique ?* Editions CEREMA, Paris.

Chabenat, G. (1996). *L'aménagement fluvial et la mémoire. Parcours d'un anthropologue sur le fleuve Rhône*. Préface de Michel Marié Paris. L'Harmattan, Paris.

Chevallier, A. (2009). Guide exercices cadre et terrain. Ministère de l'intérieur DGSCGC, Paris.

Chevallier, A. (2011). Exercices de sécurité civile, guide méthodologique sur les exercices de la sécurité civile. Ministère de l'intérieur. Gouvernement français, Paris.

Chouquer, G. (2008). *Traité d'archéogéographie. La crise des récits géohistoriques*. Editions Errance, Paris.

Clément, V. (2005). Les feux de forêt en Méditerranée : un faux procès contre Nature *L'Espace géographique*, 34(4).

Cœur, D. (2003). La maîtrise des inondations dans la plaine de Grenoble (XVIIe-XXe siècle) : enjeux techniques, politiques et urbains. History PhD Thesis, Université Grenoble 2, Grenoble.

Comfort, L.K., Siciliano, M.D., Okada, A. (2010) Risque, résilience et reconstruction : le tremblement de terre haïtien du 12 janvier 2010. *Telescope*, 16(2).

Cook, A.D.B., Shrestha, M., Htet, Z.B. (2018). An assessment of international emergency disaster response to the 2015 Nepal earthquakes. *International Journal of Disaster Risk Reduction*, 31.

Corack, L., Deschamps, S., Vinet, F. (2017). Flood crisis management: The operational perspective. In *Floods 2: Risk Management*, Vinet, F. (ed.). ISTE Ltd, London, and John Wiley & Sons, New York.

Cotelle, P. (2014). Une géographie de l'insécurité urbaine post catastrophe : le cas de la Nouvelle Orléans et du cyclone Katrina (USA 2000–2010). Geography PhD Thesis, Université Paul Valéry, Montpellier.

Cretney, R.M. (2017). Towards a critical geography of disaster recovery politics: Perspectives on crisis and hope. *Geography Compass*, 11(1). doi: 10.1111/gec3.12302.

Creton-Cazanave, L. (2010). Penser l'alerte par les distances. Entre planification et émancipation, l'exemple du processus d'alerte aux crues rapides sur le bassin versant du Vidourle. PhD Thesis, Université de Grenoble.

Crocq, L., Huberson, S., Vraie, B. (2009). *Gérer les grandes crises : sanitaires, écologiques, politiques et économiques*. Odile Jacob, Paris.

Croisé, A. (2020). L'invisibilité des cancers professionnels en Seine-Saint-Denis : déconstruire une absence sociale de risque à l'aune du recensement et des transformations des lieux d'activités délétère. Geography PhD Thesis, Université Panthéon-Sorbonne, Paris.

Croisé, A., Sierra, A., Lysaniu, B. (2018). Invisibilité des risques sanitaires et ambivalences des héritages industriels en Seine-Saint-Denis. *EchoGéo*, 46.

Crozier, D., Jouannic, G., Tran Duc Minh, C., Kolli, Z., Matagne, E., Arbizzi, S. (2017). Reconstruire un territoire moins vulnérable après une inondation. *Espace populations sociétés*, 2016(3).

Daudé, E., Grancher, D., Lavigne, F. (2021). Évacuation massive des populations en temps d'épidémie de COVID-19 : comment éviter la sur-crise? *EchoGéo*, Sur le Vif.

Dauphiné, A. (2000). *Risques et catastrophes. Observer, spatialiser, comprendre, Gérer*. A. Colin ed., Paris.

Dautun, C. and Lacroix, B. (2013). Crise et décision : plongée au cœur des cellules de crise. *Cahiers de la Sécurité*, 24.

David, G. (2004). *Espaces tropicaux et risques, du local au global*. Presses Universitaires d'Orléans, IRD, Orléans.

Davidsson, Å. (2020). Disasters as an opportunity for improved environmental conditions. *International Journal of Disaster Risk Reduction*, 48.

Dedieu, F. (2013). *Une catastrophe ordinaire. La tempête du 27 décembre 1999*. Éditions de l'EHESS, Paris.

Deffarges, T. (2003). Sur la nature et les causes du terrorisme. Une revue de la littérature économique. *Revue Tiers Monde*, 44.

Degardin, F. (2002). Urbanisation et inondations : de l'opposition à la réconciliation. *Bulletin de l'Association de Géographes Français*, 79(1).

Degardin, F. and Gaide, P.A. (1999). *Valoriser les zones inondables dans l'aménagement urbain : repères pour une nouvelle démarche*. Dossier CERTU, Paris.

Desailly, B. (1989). Les ingénieurs des Ponts et Chaussées face aux inondations en Roussillon (1770-1800). *Revue géographique des Pyrénées et du Sud-Ouest*, 60(3).

De Vanssay, B. (1999). Une étude exemplaire, une postérité bien établie : les études de perception des risques sismiques et volcaniques aux Antilles et en Métropole. In *Les catastrophes naturelles aux Antilles : d'une Soufrière à l'autre*, Yacou, A. (ed). Karthala, CERC, Paris.

De Vanssay, B. (2003). Quand les sciences humaines éclairent l'analyse des risques. *Pouvoirs locaux*, 56.

De Vanssay, B. (2010). La reconstruction, développement durable et réduction de la vulnérabilité. Face à la post-catastrophe. Institut des Risques Majeurs. *Risques Info*, 24.

Di Méo, G. (1998). *Géographie sociale et térritoires*. Editions Nathan, Paris.

Dorna, A. and Borges Costa, G. (2015). La crise politique et la perte de cohésion sociale. *Les cahiers psychologie politique*, 26.

Douvinet, J. (2020). *Alerter la population face aux crues rapides en France*. Éditions Quae, Paris.

Douvinet, J., Kouadio, J., Bonnet, E., Gensel, J. (2017). Crowdsourcing and crisis-mapping in the event of floods: Tools and challenges. In *Floods 2: Risk Management*, Vinet, F. (ed.). ISTE Ltd, London, and John Wiley & Sons, New York.

Douvinet, J., Gisclard, B., Martin, G., Vinet, F., Bopp, E., Grancher, D., Coulon, M., Genre-Grandpierre, C. (2019). Les sirènes sont-elles pertinentes en France pour alerter la population en cas de crues rapides ? *La Houille Blanche*, 105.

Drabeck, T.E. (2010). *The Human Side of Disaster*. CRC Press, Taylor & Francis, London.

Duclos, D. (1987). La construction sociale du risque : le cas des ouvriers de la chimie face aux dangers industriels. *Revue française de sociologie*, 28(1).

Dupuy, J.P. (1991). *La Panique*. Les Empêcheurs de penser en rond, Paris.

Dutozia, J. and Voiron-Canicio, C. (2018). La résilience urbaine face aux risques, en contexte de changement climatique. UNS – IMREDD, Laboratoire ESPACE, VINCI Construction France, Nice.

Duvat, V.K.E., Volto, N., Stahl, L., Moatty, A., Defossez, S., Desarthe, J., Grancher, D., Pillet, V. (2021). Understanding interlinkages between long-term trajectory of exposure and vulnerability, path dependency and cascading impacts of disasters in Saint-Martin (Caribbean). *Global Environmental Change*, 67(B).

Duvignaud, J. (1974). *La synthèse écologique*. Edition Doin, Paris.

Dynes, R.R. (1970). *Organized Behaviour in Disaster*. Heath Lexington Books, Lexington.

Dynes, R.R. (1994). Community emergency planning: False assumptions and inappropriate analogies. *International Journal of Mass Emergencies and Disasters*, 12(2).

Ecole Nationale d'Administration (2020). Gérer localement la crise du COVID 19. *Revue française d'administration publique*, 176.

d'Ercole, R. (1996). La catastrophe et son impact à moyen terme : l'exemple de Popayan (Colombie), dix ans après le séisme du 31 mars 1983. *Cahiers des Sciences Humaines*, 96–2.

d'Ercole, R. (2004). *Système d'information et risques dans le District Métropolitain de Quito. IRD: 30 ans en Equateur*. IRD, Quito.

d'Ercole, R. and Metzger, P. (2002). *Los lugares esenciales del Distrito Metropolitano de Quito*. Colección Quito Metropolitano, MDMQ-IRD, Quito, Ecuador.

d'Ercole, R. and Metzger, P. (2004). *Vulnerabilidad del Distrito Metropolitano de Quito*. Colección Quito Metropolitano, MDMQ-IRD, Quito, Ecuador.

d'Ercole, R. and Metzger, P. (2005). Repenser le concept de risque pour une gestion préventive du territoire, Rethinking the Concept of Risk for Preventive Land Management. *Pangea*, 43(44).

d'Ercole, R. and Metzger, P. (2009). La vulnérabilité territoriale : une nouvelle approche des risques en milieu urbain. *Cybergeo : Revue européenne de géographie/European Journal of Geography*, Dossier spécial.

d'Ercole, R. and Rançon, J.-P. (1994). La future éruption de la Montagne Pelée : risques et représentations. *Mappemonde*, 4.

d'Ercole, R. and Rançon, J.-P. (1999). Représentation des risques liés à une éruption future de la Montagne Pelée (Martinique) : confrontation des points de vue de populations proches (Saint-Pierre, Le Prêcheur, Le Morne-Rouge) et des volcanologues. In *Les Antilles, terres à risque*, Pagney, F. and Leone, F. (eds). *Terres d'Amérique*, 2, Géode Caraïbe Karthala, Paris.

d'Ercole, R., Thouret, J.C., Dollfus, O., Asté, J.P. (1994). Les vulnérabilités des sociétés et des espaces urbanisés : concepts, typologie, modes d'analyse. *Revue de géographie alpine*, 82(4).

d'Ercole, R., Metzger, P., Robert, J., Hardy, S., Gluski, P., Vernier, P., Sierra, A., Perfettini, H., Guillier, B. (2011). *Recursos de respuesta inmediata y de recuperación temprana ante la ocurrencia de un sismo y/o tsunami en Lima Metropolitana y Callao*. Estudio SIRAD, Proyecto Preparación ante desastre sísmico y/o tsunami y recuperación temprana en Lima y Callao, ECHO/PNUD/INDECI/COOPI/IRD, Lima.

d'Ercole, R., Hardy, S., Metzger, P., Robert, J., Gluski, P. (2012). Les dimensions spatiales et territoriales de la gestion de crise à Lima. *Vertigo*, 12(1).

Fanning, R.M. and Gaba, D. (2007). The role of debriefing in simulation based on learning. *Simulation in Healthcare*, 2(2).

Favier, R. (2007). L'histoire sociale des risques naturels en question. Sources et problématiques nouvelles. *Colloque Gestion sociale des risques naturels*, Région Autonome de la Vallée d'Aoste, sous la direction de René Favier et Claudine Remacle.

Ferretti, A. (2015). *Les territoires face aux catastrophes naturelles, quels outils pour prévenir les risques ?* Rapport du CESE, Paris.

Fink, S. (2000). *Crisis Management: Planning for the Inevitable*. Edition Iuniverse, Bloomington.

Fleming, J. and Ledogar, R.J. (2008). Resilience, an evolving concept: A review of literature relevant to aboriginal research. *Pimatisiwin*, 6(2).

Forgues, B. (1996), Nouvelles approches de la gestion de crise. *Revue Française de Gestion*, 108.

Framery, D. (2003). Les inondations de la Somme en 2001 : enjeux de pouvoirs dans un contexte de décentralisation. *Hérodote*, 3(110).

Fréalle, N. (2018). Formation à la gestion de crise à l'échelle communale : méthode d'élaboration et de mise en œuvre de scénarios de crise crédibles, pédagogiques et interactifs. PhD Thesis, Université de Lyon.

Fujiki, K. (2017). Etude prospective des impacts sociaux d'une inondation majeure en Ile-de-France. Disparités socio-spatiales dans la prise en charge des populations franciliennes en situation de crise et post-crise : une analyse cartographiée et quantifiée des besoins des ménages, de l'évacuation et de la reconstruction. Geography PhD Thesis, Université Lyon 3, Lyon.

Gaillard, J.C. (2002). Implications territoriales et ethnoculturelles d'une crise volcanique : le cas de l'éruption du Mont Pinatubo aux Philippines. *Annales de Géographie*, 111(627/628).

Gaillard, J.C. (2006). La culture comme enjeu majeur de la gestion des catastrophes liées à des phénomènes naturels au sein des sociétés traditionnelles. In *La vulnérabilité des sociétés et des territoires face aux menaces naturelles*, Leone, F. and Vinet, F. (eds). Publications de l'université Paul Valéry-Montpellier 3, Montpellier.

Gaillard, J.C. (2008). Alternative paradigms of volcanic risk perception: The case of Mt. Pinatubo in the Philippines. *Journal of Volcanology and Geothermal Research*, 172(3–4).

Gaillard, J.C. (2010a). Vulnerability, capacity and resilience: Perspectives for climate and development policy. *Journal of International Development*, 22(2).

Gaillard, J.C. (2010b). De l'utilité des légendes et toponymes dans la reconstruction et l'anticipation de catastrophes : le cas du Mont Pinatubo aux Philippines. *Quelles archives aujourd'hui pour mieux gérer les risques demain ? Approches géographiques et historiques*, Meschinet de Richemond, N. (ed.). Publications de l'Université Paul Valéry-Montpellier 3, Montpellier.

Gaillard, J.C. (2013). Livelihoods and disasters. In *Encyclopedia of Natural Hazards, Encyclopedia of Earth Sciences Series*, Bobrowsky, P.T. (eds). Springer, Dordrecht.

Gaillard, J.C. (2019). Disaster studies inside out. *Disasters*, 43(S1).

Gaillard, J.C., Wisner, B., Benouar, D., Cannon, T., Creton-Cazanave, L., Dekens, J., Fordham, M., Gilbert, C., Hewitt, K., Kelman, I. et al. (2010). Alternatives for sustained disaster risk reduction. Human Geography, *Geography and Environmental Studies*, Publications. 16–3(1).

Galland, J.-P. (2010). Critique de la notion d'infrastructure critique. *Flux*, 3(81).

Gallois, N. (2012). Les crises: L'analyse des économistes français du XIXe. *L'Économie politique*, 3(55).

Gaultier-Gaillard, S., Persin, M., Vraie, B. (2011). *Gestion de crise : les exercices de simulation : de l'apprentissage à l'alerte*. Afnor Editions, Paris.

Gavilanes-Ruiz, J.C., Cuevas-Muñiz, A., Varley, N., Gwynne, G., Stevenson, J., Saucedo-Girón, R., Pérez-Pérez, A., Aboukhalil, M., Cortés-Cortés, A. (2009). Exploring the factors that influence the perception of risk: The case of volcán de Colima, Mexico. *Journal of Volcanology and Geothermal Research*, 186(3–4).

Gerard, F.E. and Lang, M. (2019). Xynthia : analyse des causes et des conséquences de la catastrophe. *La Houille Blanche*, 3–4.

Ghosh, S., Huyck, C., Greene, M., Gill, S., Bevington, J., Svekla, W., DesRoches, R., Eguchi, R. (2011). Crowdsourcing for rapid damage assessment: The global earth observation catastrophe assessment network (GEO-CAN). *Earthquake Spectra*, 27(S1).

Gilbert, C. (1992). *Le pouvoir en situation extrême : catastrophes et politique*. L'Harmattan, Paris.

Gilbert, C. (2001). Retour d'expérience, le poids des contraintes. *Annales des Mines, Responsabilité et environnement*, 22.

Gilbert, C. (2005). L'analyse des crises : entre normalisation et évitement, in risques, crises et incertitudes : pour une analyse critique. In *Cahiers du GIS Risques Collectifs et Situations de Crise*, Borraz, O., Gilbert, C., Joly, P.B. (eds). CNRS-Publications de la MSH-Alpes, Grenoble.

Glatron, S. (2009). Représentations cognitives et spatiales des risques et des nuisances pour les citadins. HDR Geography Thesis, Université de Strasbourg.

Godard, O., Henry, C., Lagadec, P., Michel-Kerjan, E. (2002). *Traité de nouveaux risques*. Edition Gallimard, Paris.

Goeldner-Gianella, L. (2007). Dépoldériser en Europe occidentale. *Annales de géographie*, 656(4).

Goeldner-Gianella, L. (2012). Changement climatique et dépoldérisation : le rôle des acteurs et le poids des représentations sociales sur les côtes d'Europe atlantique. *Quaderni*, 41.

Goeldner-Gianella, L. (2018). La dépoldérisation dans les marais littoraux d'Europe occidentale et de France : quelles entraves sociales et quelles solutions ? *Colloque international adaptation des marais littoraux au changement climatique*, 27-29/11/18, La Rochelle.

Gralepois, M. (2012). *Face aux risques d'inondation*. Éditions d'ULM, Paris.

Grünewald, F. (2010). Les enseignements de la catastrophe [Online]. Available at: http://journals.openedition.org/humanitaire, (27).

Grünewald, F. and Renaudin, B. (2010). Etude en temps réel de la gestion de la crise en Haïti après le séisme du 12 janvier 2010. Mission du 9 au 23 février 2010, Etude financée par la Délégation aux Affaires Stratégiques du Ministère de la Défense, Paris.

Grünewald, F., Binder, A., Georges, Y. (2010). Evaluation en temps réel en Haïti : trois mois après le tremblement de terre. ALNAP, Inter Agency Standing Committee, Haiti.

Guillaumin, J. (2004). Pour une méthodologie générale des recherches sur les crises. In *Crise, rupture et dépassement*, Kaes, R. (ed.). Dunod, Paris.

Hallegatte, S. and Przyluski, V. (2010). The economics of natural disasters: Concepts and methods. Policy Research Working paper, 5507, The World Bank, New York.

Hardy, S, (2013). *Atlas de la vulnérabilité de l'agglomération de La Paz*. Éditions de l'IRD, Marseille.

Hardy, S. (2020). Las alternativas de abastecimiento en agua en la aglomeración paceña : herramienta eficiente de integración y de resiliencia de la gestión urbana del agua. *Agua Y Desigualdades Urbanas*, AFD-CIDES-UMSA, La Paz.

Hay, J. and Thebaud, O. (2002). Evaluation économique et indemnisation des dommages causés par les marées noires : enseignements tirés du cas de l'Amoco Cadiz. *Economie Appliquée*, 55(4).

der Heide, E.A. (1989). *Disaster Response: Principles of Preparation and Coordination*, Mosby, Toronto, Canada.

Henry, N. (2007). Perception du risque inondation chez les acteurs industriels situés en zone inondable. GCRN Master's Thesis, Université Paul-Valéry, Montpellier.

Herlemont-Zoritchak, N. (2009). "Droit d'ingérence" et droit humanitaire : les faux amis. *Humanitaire*, (23).

Hermann, C.F. (1963). Some consequences of crisis which limit the viability of organizations. *Administrative Science Quarterly*, 8.

Hernandez, J. (2009). The long way home: New Orleans, still a disaster area three years after Hurricane Katrina. *Espace Geographique*, 38(2).

Hernandez, J. (2010). Renew Orleans ? Résilience urbaine, mobilisation civique et création d'un "capital de reconstruction" à la Nouvelle Orléans après Katrina. Geography PhD Thesis, Université de Paris Ouest-Nanterre-La Défense.

Hill, M. and Gaillard, J. (2013). Integrating disaster risk reduction into post-disaster reconstruction: A long-term perspective of the 1931 earthquake in Napier, New Zealand. *New Zealand Geographer*, 69(2).

Holguín-Veras, J., Jaller, M., Van Wassenhove, L.N., Pérez, N., Wachtendorf, T. (2012). On the unique features of post-disaster humanitarian logistics. *Journal of Operations Management*, 30(7–8).

Huet, P. (2005). La méthodologie des retours d'expérience après les accidents naturels. Première tentative de codification. Report no. IGE/05/017, Paris.

Huret, R. (2010). *Katrina, 2005. L'ouragan, l'Etat et les pauvres aux Etats-Unis*. EHESS, Paris.

Ingram, J.C., Franco, G., Rumbaitis del Rio, C., Khazai, B. (2006). Post-disaster recovery dilemmas: Challenges in balancing short-term and long-term needs for vulnerability reduction. *Environmental Science and Policy*, 9(7–8).

IRGC (2005). Risk governance: Towards an integrative approach. White Paper, International Risk Governance Council, Geneva.

Jabouin, E. (2012). Haïti, en situation post-séisme : quelques effets de la catastrophe du 12 janvier 2010 sur la population locale. *Etudes Caribeennes*, 17.

Jeggle, T. and Boggero, M. (2018). Post-disaster needs assessment, post-disaster needs assessment. European Commission, GFDRR, UNDP, and the World Bank, Brussels-New York.

Jha, A.K. (2010). *Safer Homes, Stronger Communities, Safer Homes, Stronger Communities*. The World Bank, New York.

Kachali, H., Storsjö, I., Haavisto, I., Kovács, G. (2018). Inter-sectoral preparedness and mitigation for networked risks and cascading effects. *International Journal of Disaster Risk Reduction*, 30, Part B.

Kaniasty, K. (2020). Social support, interpersonal, and community dynamics following disasters caused by natural hazards. *Current Opinion in Psychology*, 2020(32).

Klemas, V. (2015). Remote sensing of floods and flood-prone areas: An overview. *Journal of Coastal Research*, 31(4).

Kosarzycki, M., Salas, E., Wilson, K., De Rouin, R. (2002). Teamwork training in organizations: The case of training resource management. *Psykhe*, 11.

Krüke, (2015). Planning for crisis rResponse: The case of the population contribution. In *Safety and Reliability of Complex Engineered Systems: ESREL*, Podofillini, L., Sudret, B., Stojadinovi, C.B., Zio, E., Kröger, W. (eds). Taylor and Francis Group, London.

Labeur, C. (2011). Des catastrophes et des hommes : portrait d'une sociabilité événementielle. Le quotidien des inondations dans la basse vallée du Rhône entre 1755 et 2003. Sociology Thesis, Université Aix Marseille.

La Branche, S. (2012). La schizophrénie écologique : le cas des déplacements quotidiens à Lyon". *Vertigo – la revue électronique en sciences de l'environnement*, special issue, 11.

Lacombe, E.D. and Paquot, T. (2016). Romorantin : vivre avec l'indiscipline des eaux. *Revue Ekologik*, 52.

Lagadec, P. (1981). *Le Risque technologique majeur. Politique, risque et processus de développement.* Pergamon Press, Paris.

Lagadec, P. (1991). *La gestion des crises : outils de réflexion à l'usage des décideurs, science*. McGraw Hill – Ediscience, Paris.

Lagadec, P. (1992). Le retour d'expérience, de l'analyse des risques à l'analyse des crises. *Préventique*, 44.

Lagadec, P. (1994). Situation de crise : apprentissage de la communication. *Sécurité-Revue de Préventique*, 13.

Lagadec, P. (1995). *Cellules de Crise : les conditions d'une conduite efficace*. Editions d'Organisation, Paris.

Lagadec, P. (2006). Crisis management in the twenty-first century: Unthinkable events in inconceivable contexts. In *Handbook of Disaster Research*, Rodriguez, H., Quarantelli, E.L., Dynes, R.R. (eds). Springer, New York.

Lagadec, P. (2007a). Enseigner sur la question des crises : Enjeux, Obstacles, Initiatives. Cahiers du Laboratoire d'Econométrie, 2007-07.

Lagadec, P. (2007b). Katrina : examen des rapports d'enquête. Tome 1 : A Failure of Initiative. US House of Representatives.

Lagadec, P. (2007c). Katrina : examen des rapports d'enquête. Tome 2 : The Federal Response to Hurricane Katrina – Lessons Learned. Cahiers du Laboratoire d'Econométrie, 2007-11.

Lagadec, P. (2012). *Du risque majeur aux mégachocs.* Les cahiers de Préventique/12, Editions Préventique, Paris.

Lagadec, P. and Guilhou, X. (2001). Les conditions de survenue des crises graves. Conditions et mécanismes de production des défaillances, accidents et crises. In *Actes de la seconde séance du Séminaire "Le risque de défaillance et son contrôle par les individus et les organisations dans les activités à hauts risques*", Amalberti R., Fuchs C. and Gilbert C. (eds). CNRS, Maison des Sciences de l'Homme, Paris.

Laganier, R. (eds) (2006). *Territoires, inondation et figures du risque, la prévention au prisme de l'évaluation.* L'Harmattan, Paris.

Laganier, R. (2013). Améliorer les conditions de la résilience urbaine dans un monde pluriel : des défis et une stratégie sous contrainte. *Responsabilité & Environnement*, numéro thématique la résilience : plus qu'une mode, 72/2013.

Laganier, R. (2015). Organizational resilience: Preparing and over coming crisis. In *Resilience Imperative: Uncertainty, Risks and Disasters*, Reghezza-Zitt, M. and Rufat, S. (eds). ISTE Ltd, London, and John Wiley & Sons, New York.

Lambert, M.L (2015). Le recul stratégique : de l'anticipation nécessaire aux innovations juridiques. *Vertigo – la revue électronique en sciences de l'environnement*, special edition 21.

Lambert, A., Rey, V., Provansal, M., Samat, O., Sabatier, F. (2007). Lutte contre l'érosion littorale : efficacité des méthodes de stabilisation par drainage de plage, le cas de la baie d'Agay, Var. *Méditerranée*, 108.

Lamond, J.E., Joseph, R.D., Proverbs, D.G. (2015). An exploration of factors affecting the long term psychological impact and deterioration of mental health in flooded households. *Environmental Research*, 140.

Lançon, F. and Mendez del Villar, P. (2008). La flambée des prix mondiaux du riz : crise conjoncturelle ou mutation durable. *Hérodote*, 131(4).

Langumier, J. (2015). Les territoires face aux catastrophes naturelles : quels outils pour prévenir les risques ? Etude présentée par M. Alain Feretti au nom de la Délégation à la prospective et à l'évaluation des politiques publiques présidée par Jean-Paul Bailly, Conseil Economique Social et Environnemental, Paris.

Langumier, J. and Verdier, L. (2015). Sécurisation des digues et dispositifs de culture du risque: Une alliance originale sur le delta du Rhône dans un contexte post-catastrophe. *Géocarrefour*, 90(3).

Lapierre, D. (2016). Méthode EVADE : une approche intégrée pour l'EValuation et l'Aide au Debriefing. PhD Thesis, Université de Nîmes.

Lapierre, D. and Fréalle, N. (2020). Quelles sont les limites rencontrées dans l'élaboration des exercices de formation à la gestion de crise aujourd'hui ? *Perspectives*, 21.

La Porte, T.M. (2006). Organizational strategies for complex system resilience, reliability, and adaptation. In *Seeds of Disaster, Roots of Response, How Private Action can Reduce Public Vulnerability*, Auerswald, P.E., Branscomb, L.M., La Porte, T.M., Michel-Kerjan, E.O. (eds). Cambridge University Press, New York.

Larroque, C. (2014). Corruption de la démocratie et enjeu environnemental : la crise des ordures napolitaine. *Les ateliers de l'éthique / The Ethics Forum*, 9(1).

Lavigne, F. and Thouret, J.C. (2000). Les lahars : dépôts, origines et dynamique. *Bulletin de la Société Géologique de France*, 171(5).

Lavigne, F., de Coster, B., Juvin, N., Flohic, F., Gaillard, J.C., Texier, P., Morin, J., Sartohadi, J. (2008). People's behavior in the face of volcanic hazards: Perspectives from Javanese communities, Indonesia. *Journal of Volcanology and Geothermal Research*, 172(3–4).

Le Bouëdec, G. (2011). Le temps des crises dans les sociétés littorales (xvie–xixe siècle). *Annales de Bretagne et des Pays de l'Ouest*, 117-3.

Le Gouill, C., Cortinas Munoz, J., Poupeau, F. (2019). Coupures d'eau et crise politique. Éléments pour une sociologie des transformations de l'État en Bolivie. *Politix*, 127(3).

Lecomte, M. (2008). La microfinance solidaire. *Revue d'économie financière*, 92.

Le Houërou, H.-N. (1993). Changements climatiques et désertisation. *Sécheresse*, 4(2).

Lenôtre, N. (2009). Pour une gestion dynamique du littoral. *Responsabilité et environnement*, 56.

Leone, F. (2007). Caractérisation des vulnérabilités aux catastrophes naturelles : contribution à une évaluation géographique multirisque (mouvements de terrain, séismes, tsunamis, éruptions volcaniques, cyclones). HDR, Geography, Université Paul Valéry, Montpellier.

Leone, F. and Vinet, F. (eds) (2015). *Plans communaux de sauvegarde et outils de gestion de crise*. Presses universitaires de la Méditerranée, Montpellier.

Leone, F., Meschinet de Richemond, N., Vinet, F. (2010). *Aléas naturels et gestion des risques*. PUF, Paris.

Lesquel, E. (2012). Accidents industriels. Accords et désaccords autour des plans de prévention. *La gazette des communes*, 2141–7/09.

Levy, A. (2009). La ville durable paradoxes et limites d'une doctrine d'urbanisme émergente. *Le cas Seine-Arche, Revue Esprit*, 12.

Lhomme, S. (2012). Les réseaux techniques comme vecteur de propagation des risques en milieu urbain. Une contribution théorique et pratique à l'analyse de la résilience urbaine. Geography PhD Thesis, Université Paris Diderot, Ecole d'Ingénieur de la ville de Paris, Paris.

Lhomme, S., Serre, D., Diab, Y., Laganier, R. (2010). Les réseaux techniques face aux inondations, ou comment définir des indicateurs de performance de ces réseaux pour évaluer la résilience urbaine. *Bulletin de l'Association de géographes français*, 4.

Lhomme, S., Laganier, R., Diab, Y., Serre, D. (2012). La résilience urbaine : un nouveau concept opérationnel vecteur de durabilité urbaine ? *Développement Durable et territoires*, 3(1).

Libaert, T., Baygert, N., Motulsky, B., Vanderbiest, N., Vicherat, M. (2018). *Communication de crise*. Pearson, France.

Lizarralde, G., Johnson, C., Davidson, C. (2009). *Rebuilding After Disasters: From Emergency to Sustainability*. Routledge, London.

Lumbroso, D. and Vinet, F. (2011). A comparison of the causes, effects and aftermaths of the coastal flooding of England in 1953 and France in 2010. *Natural Hazards and Earth System Sciences*, 11.

Lumbroso, D., Stone, K., Vinet, F. (2011). An assessment of flood emergency plans in England and Wales, France and the Netherlands. *Natural Hazards*, 58(1).

Lyons, M. (2009). Building back better: The large-scale impact of small-scale approaches to reconstruction. *World Development*, 37(2).

Magnan, A.K., Schipper, E.L.F., Burkett, M., Bharwani, S., Burton, I., Eriksen, S., Gemenne, F., Schaar, J., Ziervogel, G. (2016). Addressing the risk of maladaptation to climate change. *Wiley Interdisciplinary Reviews: Climate Change*, 7(5).

Mahul, O. and Ghesquiere, F. (2010). Financial protection of the state against natural disasters: A primer. Policy Research Working Paper, The World Bank, New York.

Mainguet, M. (1995). *L'homme et la sécheresse*. Masson Coll. Géographie, Paris.

Mantey, K., Coccoz, F., Boulogne, O., Torrents, R., Guibert, N., Six, C., Malfait, P. (2012). Surmortalité associée à un déplacement en urgence des personnes âgées hébergées en institution suite aux inondations du Var en 2010. *Geriatr Psychol Neuropsychiatr Vieil.*, 10(4).

Maret, I. and Cadoul, T. (2008). Résilience et reconstruction durable : que nous apprend la Nouvelle-Orléans. *Annales de géographie*, 663(5).

Martel, A. (2014). Coordination humanitaire en Haïti : le rôle des clusters dans l'externalisation de l'aide. *Mondes en développement*, 165(1).

Martin, P.M. (2012). *Crise du terrorisme, crise de l'état de droit crise(s) et droit*. Presses Université de Toulouse, Toulouse.

Mas, M. (2012). Analyse comparative des représentations du risque volcanique en milieu insulaire : Guadeloupe, Martinique et Réunion. Geography PhD Thesis, Université Paul Valéry, Montpellier.

Masclet, G. (2010). Entre chaos et entropie : une solution à la crise ? *Le journal des psychologues*, 10(283), 22–25.

Masson, E., Fischer, C., Laurens, S., Raude, J. (2003). La crise de la vache folle psychose, contestation, mémoire et amnésie. *Connexions*, 2(80).

Mathurin, E. (2012). La vulnérabilité sociale à la veille du séisme. In *Haïti, réinventer l'avenir*, Rainhorn, J.D. (ed.). Éditions de la Maison des sciences de l'homme, Paris

Mayfield, J. and Mayfield, M. (2017). Leadership communication: Reflecting, engaging, and innovating. *International Journal of Business Communication*, 54(1).

Mayfield, J., Mayfield, M., Sharbrough, W.C. (2015). Strategic vision and values in top leaders communications: Motivating language at a higher level. *International Journal of Business Communication*, 52.

McEntire, D. (2012). Understanding and reducing vulnerability: From the approach of liabilities and capabilities. *Disaster Prevention and Management*, 21(2).

MEEM (2016). La démarche française de prévention des risques majeurs. Ministère de l'Environnement de l'Energie et de la Mer, Direction Générale de la Prévention des Risques, Paris.

Melé, P. (1998). *Reconstruction des quartiers anciens racines de notre ville. Dans Patrimoine et action publique au centre des villes mexicaines*. Éditions de l'IHEAL, Paris.

Mendenhall, E. (2020). The COVID-19 syndemic is not global: Context matters. *The Lancet*, 396(10264).

Menoni, S. (2001). Chains of damages and failures a metropolitan environment: Some observations on the Kobe earthquake. *Journal of Hazardous Materials*, 86(1–3).

Meschinet de Richemond, N. (2012). Risques, crises et territoires : réflexions géographiques et historiques sur les cindyniques. HDR, Université Paul Valéry, Montpellier.

Meschinet de Richemond, N. (2016). Modernité, anachronisme et ambivalence des risques et catastrophes naturelles à travers l'approche géohistorique. *Vertigo*, 16(3).

Meschinet de Richemond, N. and Reghezza-Zitt, M. (2010). La gestion du risque en France : contre ou avec le territoire ? *Annales de géographie*, 673(3).

Metzger, P. (2009). L'épidémie de chikungunya : un problème de moustiques ? In *Comment se construisent les problèmes de santé publique*, Gilbert, C. Gilbert, C., Henry, E. (eds). La Découverte, Paris.

Metzger, P., Robert, J., Sierra, S., d'Ercole, R., Hardy, S., Gluski, P. (2013). Dimensions spatiales et territoriales de la gestion de crise : les ressources de décision et d'intervention à Lima et Callao. *Revue Géographique de l'Est*, 53(1–2).

Metzger, P., Gluski, P., Robert, J., Sierra, A. (2014). *Atlas problématique d'une métropole vulnérable. Inégalités urbaines à Lima et Callao*. Editions IRD, Bondy.

Michel-Kerjan, E. (2003a). Risques catastrophiques et réseaux vitaux : de nouvelles vulnérabilités. *Flux*, 51(1).

Michel-Kerjan, E. (2003b). Terrorisme à grande échelle partage de risques et politiques publiques. *Revue d'économie politique*, 5(113).

Miles, B. and Morse, S. (2007). The role of news media in natural disaster risk and recovery. *Ecological Economics*, 63(2–3).

Milian, J. and Tamru, B. (2018). Port-au-prince, ville du risque ? Un mythe au prisme d'une urbanisation vulnérable. *Etudes caraibéennes*, 40(39).

Mineo-Kleiner, L. and Meur-Ferec, C. (2016). Relocaliser les enjeux exposés aux risques côtiers en France : points de vue des acteurs institutionnels. *Vertigo : La Revue Électronique en Sciences del'Environnement*, 16(2).

Ministère des Solidarités et de la Santé (2019). *Retour d'expérience guide méthodologique Situations d'urgence sanitaire et exercices de simulation*. MSS, Paris.

Mitchell, J.T. (1997). Can hazard risk be communicated through a virtual experience? *Disasters*, 21(3).

Mitroff, I.I., Shrivastava, P., Udwadia, F.E. (1987). Effective crisis management. *The Academy of Management Executive* (1987–1989), 1(4).

Mitroff, I.I., Pauchant, T.C., Shrivastava, P. (1988). Conceptual and empirical issues in the development of a general theory of crisis management. *Technological Forecasting and Social Change*, 25.

Moatty, A. (2015). Pour une géographie des reconstructions post-catastrophe : risques, sociétés, territoires. Geography doctorate, Université Paul Valéry, Montpellier.

Moatty, A. (2020). Resilience and post-disaster recovery: A critical reassessment of anticipatory strategies build back better and capacity building. *Disaster Prevention and Management: An International Journal*, 29(4).

Moatty, A. and Dubos-Paillard, E. (2020). Le système d'assainissement en Ile-de-France : entre ressource et facteur aggravant pour la gestion d'une inondation majeure. *Cybergeo: European Journal of Geography*, 940.

Moatty, A. and Reghezza-Zitt, M. (2018). Infrastructures critiques, vulnérabilisation du territoire et résilience : assainissement et inondations majeures en Île-de-France. *Vertigo*, 18(3).

Moatty, A., Gaillard, J.C., Vinet, F. (2017). Du désastre au développement : les enjeux de la reconstruction post-catastrophe. *Annales de Geographie*, 714.

Moatty, A., Vinet, F., Defossez, S., Cherel, J.P., Grelot, F. (2018). Integrating a preventive ethic in the post-disaster recovery process: Resilience, adaptations and preventive reconstruction. *Houille Blanche*, 5–6.

Moatty, A., Grancher, D., Virmoux, C., Cavero, J. (2019). Bilan humain de l'ouragan Irma à Saint-Martin : la rumeur post-catastrophe comme révélateur des disparités socio-territoriales. Human toll of hurricane Irma in Saint-Martin: The post-disaster rumour reveals socio-territorial disparities. *Géocarrefour*, 93(1).

Moatty, A., Grancher, D., Virmoux, C., Cavero, J. (2020). Organisation de la post-catastrophe après Irma à Saint-Martin. *EchoGéo*, (51).

Molinari, D., Menoni, S., Ballio, F. (eds) (2017). *Flood Damage Survey and Assessment: New Insights from Research and Practice American Geophysical Union*. Wiley, New York.

Monteil, C., Simmons, P., Hicks, A. (2020). Post-disaster recovery and sociocultural change: Rethinking social capital development for the new social fabric. *International Journal of Disaster Risk Reduction*, 42.

Morin, E. (1972). Le retour de l'événement. *Communications*, 18.

Morin, E. (1976). Pour une crisologie. *Communications*, 25.

Morin, E. (1977). *La méthode : la nature de la nature*. Le Seuil, Paris.

Morin, E. (1980). *La méthode : la vie de la vie*. Le Seuil, Paris.

Morin, E. (1986). *La méthode : la connaissance de la connaissance*. Le Seuil, Paris.

Morin, E. (1994). *Sociologie*. Le Seuil, Paris.

Morin, E. and Béjin, A. (1976). *Introduction Communications*. Le Seuil, Paris.

Morsel, J. (1988). *Crise ? Quelle Crise ? Remarques à propos de la prétendue crise de la noblesse allemande à la fin du Moyen Âge*. Travaux historiques, Paris.

MTES (2017). Rapport 2015–2017 du délégué aux risques majeurs. Ministère de la transition ecologique et solidaire, DGPR, Paris.

Mucchielli, A. (2009). *Communication et influence : approche situationnelle*. Editions Ovadia, Nice.

Nathan, F. (2012). *La régulation sociale des risques de catastrophe : ethnographie des quartiers périphériques de La Paz*. Nouvelle édition, Geneva.

Nations Unies (2015). *Sendai Framework for Disaster Risk Reduction 2015–2030*. Nations Office for Disaster Risk Reduction, Sendai.

Neuilly, M.T. (2008). Katrina, une catastrophe exemplaire. In *Gestion et prévention de crise en situation post-catastrophe*. De Boeck Supérieur, Paris.

Nigg, J. (1995). Disaster recovery as a social process. Preliminary Paper 219, Disaster Research Center.

November, V. (2002). *Les territoires du risque : le risque comme objet géographique*. Peter Lang, Bern.

November, V. and Créton-Cazanave, L. (2017). La gestion de crise à l'épreuve de l'exercice EU SEQUANA. French document, Paris.

November, V., Azémar, A., Lecacheux, S., Winter, T. (2020). Le couple anticipation/décision aux prises avec l'exceptionnel, l'imprévu et l'incertitude. *EchoGéo*, 51.

Ocqueteau, F. (2007). La sécurité globale une réponse à la menace terroriste ? Dossier : l'etat face aux risques. Regards sur l'actualité, French document, 328, Paris.

Ogrizek, M. and Guillery, J.M. (1997). *La communication de crise*. Presses Universitaires de France, Paris.

O'Keefe, P., Westgate, K., Wisner, B. (1976). Taking the naturalness out of natural disasters. *Nature*, 260.

Oliver-Smith, A. (1991). Successes and failures in post-disaster resettlement. *Disasters*, 15(1).

Oliver-Smith, A. (2020). Hurricanes, climate change, and the social construction of risk. *International Journal of Mass Emergencies and Disasters*, 38(1).

Olshansky, R.B., Hopkins, L.D., Johnson, L.A. (2012). Disaster and recovery: Processes compressed in time. *Natural Hazards Review*, 13(3).

Orillard, F., Gralepois, M., Verdelli, L. (2018). La prévention des inondations dans les opérations d'aménagement des interfaces ville-port, un levier de gentrification indirecte ? Le cas du Havre (France). *Carnets de Géographes*, 11(11).

Özdamar, L. and Alp Ertem, M. (2015). Models, solutions and enabling technologies in humanitarian logistics. *European Journal of Operational Research*, 244(1).

Parrinello, G. (2016). Les enjeux de l'après : vulnérabilité et résilience à l'épreuve des politiques de la catastrophe au 20e siècle. *Vertigo – la revue électronique en sciences de l'environnement*, 16(3).

Pasteur, K. (2011). *From Vulnerability to Resilience: A Framework for Analysis and Action to Build Community Resilience*. Practical Action Publishing, Rugby.

Pauchant, T.C. (1988). An annotated bibliography in crisis management. *Cahier de Recherche*, Quebec, HEC, Montreal.

Pauchant, T.C. and Douville, R. (1993). Recent research in crisis management. A study of 24 authors publications from 1986 to 1991. *Industrial and Environmental Crisis Quarterly*, 7(1).

Pauchant, T.C. and Mitroff, I.I. (1992). *Transforming the Crisis Prone Organization: Preventing Individual, Organizational and Environmental Tragedies*. Jossey-Bass, San Francisco.

Paul, J. and MacDonald, L. (2016). Optimal location, capacity and timing of stockpiles for improved hurricane preparedness. *International Journal of Production Economics*, 174.

PCCIP (1997). Critical foundations Protecting America's infrastructures. The report of the president's commission on critical infrastructure protection, Washington.

Peeters, P., De Beukelaer-Dossche, M., De Wolf, P., Verwaest, T. (2013). Introduction à la protection contre les submersions en flandre flanders hydraulics research, Anvers. In *Digues maritimes et fluviales de protection contre les submersions 2e colloque national*, Royet, P. and Bonelli, S. (eds). Lavoisier, Paris.

Peinturier, E. (2020). Dans le vent des ouragans : gestion de crise et mobilités. Le cas de la Nouvelle-Orléans et sa région au cours des dernières décennies. Geography PhD Thesis, Université Paul Valéry-Montpellier 3, Montpellier.

Penning-Rowsell, E. and Wilson, T. (2006). Gauging the impact of natural hazards: The pattern and cost of emergency response during flood events. In *Journal Compilation*, Society, R.G. (ed.). 31(2).

Peretti-Watel, P. (2001). La crise de la vache folle, une épidémie fantôme ? *Sciences sociales et santé*, 19(1).

Peretti-Watel, P. (2003). Risque et innovation : un point de vue sociologique. *Innovations*, 2(18).

Peroche, M. (2016). La prise en compte du risque humain dans la gestion de crise submersion marine : du diagnostic de vulnérabilité à la modélisation des évacuations. Thesis, Université Paul-Valéry, Montpellier.

Perrow, C. (1984). *Normal Accidents: Living with High-Risk Technologies*. Basic Books, New York.

Philippe, R., Jonas, M., Antoine, M. (2007). *Les limites de l'aide humanitaire*. Infosciences – Epfl, Lausanne.

Piatyszech, E., Tardy, A., Lesbats, M., Cruz, A.M. (2017a). Natech events triggered by floods: When floods cause technological accidents. In *Floods 1: Risk Knowledge*, Vinet, F. (ed.). ISTE Ltd, London, and John Wiley & Sons, New York.

Piatyszeck, E., Ayral, P.A., Gralepois, M. (2017b). Local crisis management - the community safety plan: Challenges and obstacles to operationality. In *Floods 2: Risk Management*, Vinet, F. (ed.). ISTE Ltd, London, and John Wiley & Sons Inc.

Pigeon, P. (2012). Apports de la résilience à la géographie des risques : l'exemple de La Faute-sur-Mer (Vendée, France). *Vertigo – la revue électronique en sciences de l'environnement*, 12(1).

Plas, H. and Sierra, A. (2018). Les moto-taxi de Lima, entre relégation et ressource. *Mobilités marginales, mobilités dans les marges, marges mobiles*, Marne-la-Vallée, November.

Plein, C. (2019). Resilience, adaptation, and inertia: Lessons from disaster recovery in a time of climate change. *Social Science Quarterly*, 100(7).

Poursoulis, G., Dalongeville, R., Helly, B. (2000). Destruction des édifices minoens et sismicité récurrente en Crète (Grèce). *Géomorphologie : relief, processus, environnement*, 6–4.

PREDES (2009). *Diseño de escenario sobre el impacto de un sismo de gran magnitud en Lima Metropolitana y Callao*, Lima, INDECI, COSUDE, Peru.

Provitolo, D. (2005). Un exemple d'effets dominos : la panique dans les catastrophes urbaines. *Cybergeo, European Journal of Geography*, Systèmes, Modélisation, Géostatistiques, Document, 328.

Quarantelli, E.L. (1984). Emergent citizen groups in disaster preparedness and recovery activities. Final Project Report #33, Disaster Research Center, University of Delaware.

Quenault, B. (2013). Retour critique sur la mobilisation du concept de résilience en lien avec l'adaptation des systèmes urbains au changement climatique. *EchoGéo*, 24.

Quenault, B. (2014). Résurgence/convergence du triptyque catastrophe-résilience-adaptation pour (re)penser la fabrique urbaine face aux risques climatiques. *Développement durable et territoires*, 5(3).

Quenault, B. (2015). De Hyōgo à Sendai, la résilience comme impératif d'adaptation aux risques de catastrophe : nouvelle valeur universelle ou gouvernement par la catastrophe ? *Développement durable et territoires*, 6(3).

Quenault, B. (2016). La rhétorique de la résilience, une lueur d'espoir à l'ère de l'anthropocène ? Vers un changement de paradigme fondé sur l'acceptation de la catastrophe. *Les villes à la croisée des stratégies globales et locales des enjeux climatiques*. Presses universitaires de Laval, Laval.

Rainhor, J.-D. (2012). *Haïti, réinventer l'avenir*. Éditions de la Maison des sciences de l'homme, Paris.

Rasmussen, A (2007). Dans l'urgence et le secret. Conflits et consensus autour de la grippe espagnole, 1918–1919. *Mil neuf cent. Revue d'histoire intellectuelle*, 25, 1.

Rebotier, J. (2008). Les territorialités du risque urbain à Caracas. Les implications d'un construit socio-spatial dans une métropole d'Amérique latine. Institut des Hautes Etudes d'Amérique Latine. Geography PhD Thesis, Sorbonne Nouvelle, Paris.

Reghezza-Zitt, M. (2005). La ville, un territoire du risque privilégié ? Quand la représentation est un facteur de vulnérabilité. *Bulletin de l'association de géographes français*, 82(1).

Reghezza-Zitt, M. (2015). De l'avènement du monde à celui de la planète : le basculement de la société du risque à la société de l'incertitude. HDR, Université Paris 1 Panthéon-Sorbonne, Paris.

Reghezza-Zitt, M. (2019). Gestion de crise et incertitude(s) ou comment planifier le hors-cadre et l'inimaginable. Application aux crises résultant de crues majeures en Île-de-France. *Annales de géographie*, 726(2).

Reghezza-Zitt, M. and Rufat, S. (2015). *Resilience Imperative: Uncertainty, Risks and Disasters*. ISTE Ltd, London, and John Wiley & Sons, New York.

Renn, O., Klinke, A., Schweizer, P.J. (2018). Risk governance: Application to urban challenges. *International Journal of Disaster Risk Science*, 9(4).

Revault d'Allones, M. (2013). Crise et modernité. *Zeitschrift für politisches Denken/Journal of Political Thinking*, 7(1).

Revet, S. (2007). *Anthropologie d'une catastrophe : les coulées de boue de 1999 au Venezuela*. Presses Sorbonne-Nouvelles, Paris.

Revet, S. (2009). Les organisations internationales et la gestion des risques et des catastrophes naturels. Les Etudes du CERI, 157. *CERI, Sciences Po*, Paris.

Revet, S. (2011). Penser et affronter les désastres : un panorama des recherches en sciences sociales et des politiques internationales. *Critique Internationale*, 52(3).

Revet, S. (2012). Catastrophes naturelles et reconstructions locales en Amérique latine. Dossier du CERI, Sciences Po, Paris.

Revet, S. (2017). Géographies de la catastrophe. Ordre et désordres dans la gestion des catastrophes naturelles. *Papeles del CEIC. International Journal on Collective Identity Research*, 2017/2.

Revet, S. (2019). Témoigner au procès de la catastrophe Xynthia. Dimensions juridiques et morales de la parole des victimes. *Droit et société*, 2(102).

Revet, S. and Langumier, J. (2013). *Le gouvernement des catastrophes*. Karthala Edition, Paris.

Rey, B. (2015). Résilience systémique d'un territoire composé d'activités essentielles suite à une perturbation majeure – Approches systémique et spatiale. Thesis, Ecole des Mines de Saint Etienne.

Rey, T. and Defossez, S. (2019). Retours d'expériences post-catastrophes naturelles. *Géorisques*, 8, Collection de l'U.M.R. GRED, Montpellier.

Robert, J. (2012). Pour une géographie de la gestion de crise : de l'accessibilité aux soins d'urgence à la vulnérabilité du territoire à Lima. Geography PhD Thesis, Université de Grenoble-Chambéry, Grenoble.

Rohrbasser, J.-M. (2010). Le tremblement de terre de Lisbonne : un mal pour un bien *Annales de démographie historique*, 2(120).

Rieu, A.M. (2012). Penser après Fukushima. Technologie souveraine et contrôle démocratique. *Revue Ebisu*, 47.

Rieu, A.M. (2014). Fukushima Catastrophe: The challenge of complexity (collective reflexivity, adaptive knowledge, political innovation). In *Encyclopedia of Complexity and Systems Science*, Meyers, R. (ed.). Springer, New York.

Riou, E. (2019). L'encadrement juridique de l'Habitat flottant. Mémoire M2. GRIDAUH, Faculté de droit, Université de Montpellier.

Robert, J. (2012). Pour une géographie de la gestion de crise : de l'accessibilité aux soins d'urgence à la vulnérabilité du territoire à Lima. PhD Thesis, Université de Grenoble.

Robert, J. and d'Ercole, R. (2014). L'occupation et l'utilisation du sol en période de crise. *Cybergeo: European Journal of Geography*, Document 698.

Rode, S. and Gralepois, M. (2017). Towards an urban design adapted to flood risk? In *Floods 2: Risk Management*, Vinet, F. (ed.). ISTE Ltd, London, and John Wiley & Sons, New York.

Rode, S., Gralepois, M., Daniel-Lacombe, E. (2018). Les transactions entre la ville et l'inondation pour un urbanisme plus résilient. *La Houille Blanche – Revue Internationale de l'eau*.

Rodolpho, S.K. (1995). *Pinatubo and the Politics of Lahar: Eruption and Aftermath, 1991*. UP Press, Quezon City.

Rodríguez-Espíndola, O., Albores, P., Brewster, C. (2018). Disaster preparedness in humanitarian logistics: A collaborative approach for resource management in floods. *European Journal of Operational Research*, 264(3).

Roose, E., Sabir, M., Arabi, M., Morsli, B., Mazour, M. (2012). Soixante années de recherches en coopération sur l'érosion hydrique et la lutte antiérosive au Maghreb. *Physio-geo*, 6.

Roux-Dufort, C. (2003). *Gérer et décider en situation de crise*. Dunod, Paris.

Roux-Dufort, C. (2004). *La gestion de crise : un enjeu stratégique pour les organisations*. De Boeck, Paris.

Roux-Dufort, C. (2005). Comment en est-on arrivé là ? Du terrain de crise à la catastrophe. *Revue Spaces*, 85.

Royet, P. and Bonelli, S. (2013). *Digues maritimes et fluviales de protection contre les submersions*. Lavoisier, Paris.

Ruin, I. (2007). Conduite à contre-courant. Les pratiques de mobilité dans le Gard : facteur de vulnérabilité aux crues rapides. PhD Thesis, Université Grenoble 1 Joseph Fourier, Grenoble.

Salagnac, J.L., Marchand, D., Florence, C., Delpech, P., Axès, J.M. (2014). Impacts des inondations sur le cadre bâti et ses usagers. Final report, CSTB, Champs-sur-Marne.

Sampognaro, R. and Sicsic, M. (2012). Impact du séisme au Japon sur l'économie mondiale, lettre n°100 Trésor-eco. Ministère de l'Économie, des Finances et de l'Industrie, Paris.

Sarant, P.M., De Vanssay, B., Leone, F., Asselin de Beauville, C., Pontikis, C., Pagney, F., Koussoula-Bonneton, A., Colbeau-Justin, L., Schick, O., Lesales, T. et al. (2003). Retour d'expérience sur la prise de décision et le jeu d'acteurs : le cas du cyclone Lenny dans les Petites Antilles au regard du passé – Rapport final du programme evaluation et prise en compte des risques naturels et technologiques. Ministère de l'Ecologie et du Développement Durable (M.E.D.D.), Paris.

Schmitt, L., Lebeau, M., Trémolières, M., Defraeye, S., Coli, C., Denny, E., Dillinger, M., Beck, T., Dor, J.Ch., Gombert, Ph. et al. (2008). Le polder d'Erstein : objectifs, aménagements et retour d'expérience sur cinq ans de fonctionnement et de suivi scientifique environnemental (Rhin, France). *Ingenieries* spécial.

Scoccimarro, R. (2013). L'archipel disloqué le Japon après le 11 mars 2011. *Outre-Terre*, 35–36(1).

Shabou, S. (2016). Extrêmes hydro-météorologiques et exposition sur les routes : contribution à MobRISK : modèle de simulation de l'exposition des mobilités quotidiennes aux crues rapides. Geography PhD Thesis, Université Grenoble 1 Joseph Fourier, Grenoble.

Sierra, A. (2015). La capitale-risque ou comment le statut de capitale participe à construire le risque. *Géocarrefour*, 90(2).

Sierra, A. (2016). Définir les confins urbains par l'informalité politique : les arrangements citadins entre conflits municipaux et production de territoires indéterminés à Lima (Pérou). *L'Espace Politique*, 29(2).

Signorelli, A. (1992). Catastrophes naturelles et réponses culturelles. *Terrain* [Online], Available at : http://journals.openedition.org/terrain/3052.

Smith, G.P. and Wenger, D. (2007). Sustainable disaster recovery: Operationalizing an existing agenda. In *Handbook of Disaster Research. Handbooks of Sociology and Social Research*, Rodríguez, H., Quarantelli, E.L., Dynes, R.R. (eds). Springer, New York.

Smouts, M.-C. (2002). Un trou noir dans la mondialisation : le risque environnemental global. *VIIe congrès de l'Association Française de Science Politique* (AFSP), 18–21 September 2002, Lille.

Stauffer, T., Sands, N.P., Strobhar, D. (2017). Plug the holes in the Swiss cheese model, chemical engineering progress (CEP). *Spring Meeting and 13th Global Congress on Process Safety*, American Institute of Chemical Engineers (AIChe), San Antonio.

Staupe-Delgado, R. and Kruke, B.I. (2017). Preparedness: Unpacking and clarifying the concept. *Journal of Contingencies and Crisis Management*, 26(2).

Steinle-Feuerbach, M.F. (1995). La catastrophe et la règles des trois unités de temps, de lieu et d'action. *LPA*, 90.

Su, Y. and Le Dé, L. (2020). Whose views matter in post-disaster recovery? A case study of build back better, in Tacloban City after Typhoon Haiyan. *International Journal of Disaster Risk Reduction*, 51.

Supiot, A. (2015). *La gouvernance par les nombres*. Fayard, Paris.

Taleb, N.N. (2010). *Le cygne noir : la puissance de l'imprévisible*. Les Belles Lettres, Paris.

Tena-Chollet, F. (2012). Élaboration d'un environnement semi-virtuel de formation à la gestion stratégique de crise, basé sur la simulation multi-agents. PhD Thesis, Ecole Nationale Supérieure des Mines de Saint-Etienne, Saint-Etienne.

Terti, G. (2017). Forecasting of flash-flood human impacts integrating the social vulnerability dynamics. PhD Thesis, Université Grenoble Alpes, Grenoble.

Texier, P. (2007). Les inondations de février 2007 dans les kampung pauvres de Jakarta. *EchoGéo* [Online]. Available at: http://journals.openedition.org/echogeo/905.

Theys, J. (2002). L'approche territoriale du développement durable : condition d'une prise en compte de sa dimension sociale. *Développement durable et territoire* [Online]. Available at: http://developpementdurable.revues.org/document1475.html.

Theys, J. (2009). Scénarios pour une ville post-carbone. *Constructif*, 23 July.

Thiétart, R.A. and Forgues, B. (1997). Action, structure and chaos. *Organization Studies*, 18(1).

Thom, R. (1976). Crise et catastrophe. *Communication, La notion de crise*, 25. Le Seuil, Paris.

Tipple, G. (2005). Housing and urban vulnerability in rapidly-developing cities. *Journal of Contingencies and Crisis Management*, 13(2).

Tirex (2021). Guide de synthèse des principaux résultats du projet TIREX. Les connaissances et apprentissages de territoires cyclonés (Petites Antilles, saison cyclonique 2017) : fiches de synthèse des principaux résultats du projet ANR TIREX à l'attention des décideurs [Online]. Available at: https://fr.calameo.com/read/006591530b5bdb6016ed1?page=1

Toubin, M. (2014). Améliorer la résilience urbaine par un diagnostic collaboratif. L'exemple des services urbains parisiens face à l'inondation. Geography PhD Thesis, Université Paris Diderot, Paris.

Toubin, M., Lhomme, S., Diab, Y., Serre, D., Laganier, R. (2012). La résilience urbaine : un nouveau concept opérationnel vecteur de durabilité urbaine ? *Développement durable et territoires*, 3(1).

UN-DESA (2019). World Urbanization Prospects, the 2018 Revision. United Nations, Department of Economic and Social Affairs, Population Division, New York.

United Nations Office of Disaster Risk Reduction (UNISDR) (2015). Sendai framework for disaster risk reduction 2015–2030. Paper Presented at the *Third UN World Conference*, Sendai, Japan.

Valette, J.-F. (2018). Regards sur le séisme de Mexico de septembre 2017 : permanences et décalages par rapport à la catastrophe de 1985. *EchoGéo* [Online]. Available at: http://journals.openedition.org/echogeo/15296

Vandecasteele, I. and Lebebvre, A. (2006). De la fragilisation à la rupture du lien social : approche clinique des impacts psychiques de la précarité et du processus d'exclusion sociale. *Cahiers de psychologie clinique*, 1(26).

Vareilles, T. (2001). *Encyclopédie du terrorisme international.* L'Harmattan, Paris.

Verspieren, S. (2018). Le recul stratégique comme méthode de lutte face à l'érosion du litttoral. End of Study Thesis, Université de Liège [Online]. Available at: http://hdl.handle.net/2268.2/5039.

Veyret, Y. (2017). Saint Martin une catastrophe inévitable. *Diploweb.*

Veyret, Y. and Bouchon, S. (2002). Les risques liés aux carrières souterraines. Diversité des réponses et aménagement en Île-de-France. *Annales de géographie*, 626.

Veyret, Y. and Laganier, R. (2013). *Atlas des risques naturels en France*. Autrement, Paris.

Vinet, F. (2003). *Crues et inondations dans la France méditerranéenne. Les crues torrentielles des 12 et 13 novembre 1999 (Aude, Tarn, Pyrénées Orientales et Hérault).* Editeurs du Temps, Nantes.

Vinet, F., Gaillard, J.C., Leone, F., Denain, J.C., Clavé, E., Bachri, S. (2011a). Modalités spatiales et enjeux préventifs de la reconstruction post-tsunami à Banda Aceh. In *Tsunarisque : le tsunami du 26 décembre 2004 à Aceh, Indonésie*, Lavigne, F. and Paris, R. (eds). Editions de la Sorbonne, Paris.

Vinet, F., Gaillard, J.C., Leone, F., Denain, C., Clavé, E., Bachri, S. (2011b). Enjeux et modalités spatiales de la reconstruction post-tsunami à Banda Aceh. In *Tsunarisque : le tsunami du 26 décembre 2004 à Aceh, Indonésie*, Lavigne, F. and Paris, R. (eds). Editions de la Sorbonne, Paris.

Vinet, F., Defossez, S., Rey, T., Boissier, L. (2012). Le processus de production du risque submersion marine en zone littorale : l'exemple des territoires Xynthia. *Norois* 222/2012-1.

Virol, M. (2008). Du terrain à la carte : les ingénieurs du roi Louis XIV entre exigences et réalisations. In *Les usages des cartes (XVIIe-XIXe siècle) : pour une approche pragmatique des productions cartographiques*, Laboulais, I. (ed.). Presses universitaires de Strasbourg, Strasbourg.

Vogel, J.P. (2019). Les feux de forêts : l'impérieuse nécessité de renforcer les moyens de lutte face à un risque susceptible de s'aggraver. Information report, 739, Sénat, Paris.

Weiss, K. and Rateau, P. (2018). *Psychologie sociale et environnementale.* Editions in Press, Paris.

Weiss, L., Colbeau-Justin, L., Laurent, A., Chahraoui, K., de Vanssay, B. (2008). Aspects psychosociaux des stratégies de faire face à une inondation : soutien social, comparaison sociale et rumeur. *Revue québécoise de psychologie*, Department of Psychology, Université du Québec à Montréal, 29(1).

White, G.F. (1974). *Natural Hazards: Local, National, Global.* Oxford University Press, New York.

Whittle, R., Walker, M., Medd, W., Mort, M. (2012). Flood of emotions: Emotional work and long-term disaster recovery. *Emotion, Space and Society*, 5(1),

WHO (2017). Manuel d'exercices de simulation. World Health Organisation, Geneva.

WHO (2018). After action review and simulation exercices. World Health Organisation, Geneva.

Wievorka, M. (2009). La sociologie et la crise. Quelle crise, et quelle sociologie ? *Cahiers internationaux de sociologie*, 2(127).

Wildavsky, A. (1991). *Searching for Safety*. Transaction Publishers, New Jersey, New Brunswick.

Wisner, B., Cannon, T., Davis, I., Blaikie, P. (2004). *At Risk: Natural Hazards, People's Vulnerability and Disasters*, 2nd edition. Routledge, London.

Wisner, B., Gaillard, J.C., Kelman, I. (2012). Framing disaster: Theories and stories seeking to understand hazards, vulnerability and risk. *Handbook of Hazards and Disaster Risk Reduction*. Routledge, London.

Wulan Mei, E.T. (2013). Gestion des évacuations lors des crises volcaniques : étude de cas du volcan Merapi, Java, Indonésie. Geography PhD Thesis, Université Paris 1 Panthéon-Sorbonne, Paris.

Wybo, J.-L. (2009). Le retour d'expérience : un processus d'acquisition de connaissances et d'apprentissage. *Economica*, 19.

Wybo, J.-L. and Van Wassenhove, W. (2009). *Retours d'expérience et maîtrise des risques : pratiques et méthodes de mise en œuvre*. Editions Tec & Doc, Paris.

Xiao, Y., Cauberghe, V., Hudders, L. (2018). Humour as a double-edged sword in response to crises versus rumours: The effectiveness of humorously framed crisis response messages on social media. *Journal Contingencies and Crisis Management*, 26.

Zamoun, K. and Gorpe, T.S. (2018). Crisis management: A historical and conceptual approach for a better understanding of today's crises. In *Crisis Management Theory and Practice*, Tolla, K., Ristvej, J., Titko, M. (eds). IntechOpen.

Zaninetti, J.-M. (2013). Catastrophes et adaptations sur le littoral du Mississippi. *Annales de géographie*, 692(4).

List of Authors

Alain CHEVALLIER
Université de Montpellier III
France

Pauline GLUSKI
Institut de Recherche pour
le Développement (IRD)
France

Sébastien HARDY
Institut de Recherche pour
le Développement (IRD)
France

Richard LAGANIER
Université de Paris
France

Hoilid LAMSSALAK
RisCrises
Alès
France

Dimitri LAPIERRE
RisCrises
Alès
and
CERISC
Aix-en-Provence
France

Anne-Marie LEVRAUT
Conseil Général de l'Environnement
et du Développement Durable
France

Nancy MESCHINET DE RICHEMOND
Université de Montpellier III
France

Pascale METZGER
Institut de Recherche pour
le Développement (IRD)
France

Annabelle MOATTY
UMR RECOVER
INRAE - Aix-Marseille Université
France

Jérémy ROBERT
Université Rennes 2
France
and
Institut Français d'Etudes Andines
Lima
Peru

Samuel RUFAT
Institut Universitaire de France
Paris
France

Alexis SIERRA
Institut de Recherche pour
le Développement (IRD)
France

Yvette VEYRET
Université Paris Ouest Nanterre
France

Freddy VINET
Université de Montpellier III
France

Index

A, C

actors, 32, 35–37, 41, 44, 45, 47–56
adaptation, 110, 111, 115, 116, 128, 166, 168, 182, 184–189, 198, 200, 203
alert, 34, 41, 43–45
anticipation, 31–33, 37, 38, 40, 48, 49, 54, 59, 60, 79, 166, 195
capacity to mobilize, 93
catastrophe, 5, 6, 8, 9
common culture, 48, 148
communication, 34, 41, 45, 50
complexity, 132–134, 138, 141, 146, 151, 156
controversy, 46
coordination, 59, 74, 77, 111, 115, 138, 139, 143, 160
crisis, 2–4, 6, 8, 10, 13, 16, 17, 20–25, 27, 28, 30
  areas, 34
  cell, 58, 75, 77, 78
  environmental, 5, 21
  food, 23
  global, 5, 27
  health, 4, 21–23
  localized, 16, 17
  mega-, 5, 27–29
  technological, 4, 17
  management territories, 35
critical infrastructure, 82, 94, 95, 98

D, E

depoldering, 174, 180–182
disaster risk reduction (DRR), 32, 110, 119, 120
domino effects, 27, 30
eco-neighborhoods, 166, 189, 196–198
evacuation, 37, 40, 47, 51, 52, 54, 55
exercises, 58, 60, 61, 69–71, 74–76, 79

F, G

feedback, 58, 60–69, 71, 73, 79
  hot, 62, 63
  internal, 63, 65
  open, 66
global approach, 168, 196
governance, 110–112, 114, 117, 128

I, L

inequalities, 111, 126–128
inertia, 131, 133, 137, 145–147, 149, 154, 156, 158, 161, 162
information, 34, 37, 39, 41–44, 46–48, 50, 53, 56
learning, 57, 60, 61, 63, 66, 70, 71, 74, 76, 79

M, N, O

misinformation, 46

nuclear disasters, 17
obstacles, 136–138, 147

P, R

post-disaster recovery, 109–111, 113, 115, 122, 126, 128
preparation, 31, 32, 37–40, 45, 48–51, 54, 55
prevention, 32, 33, 35, 37–39, 41, 48–50, 55
project, 166–168, 170, 171, 174, 177, 188, 190, 191, 193, 196, 198, 200, 202, 203
protection, 167–171, 174, 175, 177, 178, 180–182, 185, 187, 196, 199, 200, 202
reflex card, 58
regulations, 115
relocation, 122, 124, 125, 200, 202, 203
representations, 132, 133, 135, 137, 141, 142, 145–147, 149, 151, 152, 155, 160, 163
resilience, 166, 168, 179, 189, 191, 192, 194, 196–200
resources
  financial, 117
  human and material, 88, 94, 105
  immaterial, 89
  land, 122
  major, 99, 100, 102, 104, 106
  support, 91
risk territory, 168, 198
rumors, 141, 142

S, T

scenario, 38, 41, 47, 48, 51, 55
shortcomings, 140, 142, 143
simulation, 58, 60, 61, 69, 70, 72, 74–79
smart city, 198, 199
solidarity, 111, 114, 117–122, 125, 127
strategic retreat, 168, 200
territorial
  renewal, 166
  restructuring, 122, 125
territorialization, 132, 133, 137, 141, 151, 154, 155, 163
training, 57, 58, 60, 69–71, 75–77, 79
transparency, 69

U, V

uncertainty, 57, 76
unexpected, 30, 31, 58
utopia, 168, 198, 199
vulnerabilities, 110, 116, 126, 128

Printed and bound by CPI Group (UK) Ltd, Croydon, CR0 4YY
13/09/2026
14960996-0001